Year Million

Atlas & Co.
New York

Year Million

Science at the Far Edge of Knowledge

Edited by
Damien Broderick

Atlas & Co. *Publishers*
15 West 26th Street, 2nd floor
New York, NY 10010
www.atlasandco.com

Distributed to the trade by W. W. Norton & Company

Printed in the United States

Interior design by Yoshiki Waterhouse
Typesetting by Maria Torres

Atlas & Co. books may be purchased for educational, business, or sales promotional use. For information, please write to info@atlasandco.com.

Library of Congress Cataloging-in-Publication Data is available upon request.

Trade paperback ISBN: 978-0-9777433-4-6
Library cloth ISBN: 978-1-934633-05-2

13 12 11 10 09 08 1 2 3 4 5 6

To the blessed memory of
H. G. Wells,
and to all humankind
through Year Million and beyond:
his children

Contents

The Mind/Body in Year Million

Into the Very Deepest Future

Introduction

Damien Broderick

> My own view is that we will successfully negotiate the hazards threatening our species. We will not kill ourselves off. We will not die off from disease. We will wax and wane with all manner of climate changes, asteroid impacts, runaway technology, and evil robots. We will persevere. . . . Perhaps this view that we are unkillable—at least as a species—is naive. But even if we are to live as long as an average *mammalian* species—between 1 and 3 million years—we still have huge stretches of time left, for our species is barely a quarter of a million years old. And who says we are average? My bet is that we will stick around until the very end of planetary habitability for this already old Earth.
>
> Peter Ward, *Future Evolution: An Illuminated History of Life to Come* [1]

A million years—it's a haunting number, quite terrifying if you put your imagination to work trying to grasp what it means, what it implies.

Casting our minds a million years into the past, we find in the wide world no trace anywhere of familiar, comforting intelligence. Yes, there are scattered hominids, *Homo antecessor*, *Homo heidelbergensis,* our ancient precursors—but perhaps they have only rudimentary speech and song, few tools, little in the way of clothing (although at least they own fire to warm them in the night), maybe no shelter from the rain and snow other than huddling beneath bushes or in caves as lightning cracks the sky. We imagine our first true ancestors to have been nasty, brutish and, if not short, then certainly short-lived. It is eerie to consider that these protohumans were not entirely alone, as we are; for they shared their planet with equally brutish, cousins—*Homo erectus*, *Homo ergaster*, others—who separated from common ancestors a quarter of a million years earlier.

In our future, we can anticipate a further splitting of our hereditary line, but on a vaster and stranger scale. A million years hence, if any of our lineage survive, they will be very different from how we are today, far more alien to us than we are to early *Homo*.

Is there any way to get a sense of those pitiless gulfs of time, a million years echoing away to either side of us? Not really, not to any degree that our local, earthbound imagination may easily capture, as we handily grasp the size of a football field or a cup of water or a skyscraper or an ant or a decade or the time remaining to shop for Christmas. We can try, though, by analogy.

You stand on a long, empty highway, left arm clasping the right hand of the person beside you, your mother or father, fingers interlaced, he or she in turn linking you to a grandparent, and on, and on, each pair of arms marking out one meter of the ground stood upon (call it a yard if you remain uneasy with the metric system). Why, then, four of you can represent a century of time. That's because roughly twenty-five years comprises a generation—the average time it takes from birth to maturity and reproduction, baby to parent. So one thousand years is forty generations, forty people linking their hands, left in right, right in left, marking out forty meters of the Earth's surface.

By the standard of ordinary experience and memory, one thousand years is a great period of time. Within a millennium, nations and empires rise and fall, tongues mutate into incomprehensibility. (Have you tried reading Old English from 1000 CE? *Beowulf*, say? Not even all the letters are the same.) Multiply that span a further thousandfold and the impact becomes overwhelming.

Now, rather than the forty people linked in a chain as long as a city block, 40,000 people extend along the highway, forty kilometers of them—twenty-five miles—much farther than the eye can see. To walk in a single day from end to end of this stretched-out township would take more endurance than most of us could muster. Yet if this linear daisy chain were composed of generations of our ancestors, we could reach the dawn of human prehistory in less than a single kilometer. Back and back in time from that point we'd trudge, our hand-clasped predecessors growing less and less familiar in appearance and behavior with every kilometer; in truth, less and less human.

That's the million years of humans and hominids stretching to our left. Extend it to the right by a further thousandfold to a billion years (if that were possible, if the Sun were not fated to burn away the skin of the world first) and the generations will stand hand-in-hand, by an uncanny coincidence, all the way around the equator.

Unless we exterminate ourselves very soon, and much of the planet with us by global heating, by pandemics—natural or engineered—spread by world-girdling travelers and freight, by the undissipated threat of thermonuclear war and other dreadful weapons, there's no reason to suppose that the next million years will witness any less change than the first million.

If anything, as proved by the accelerating rates of change during the last couple of centuries, and especially the last few decades, the future will be far stranger and more devious than the past. If the gulf of time and difference between us and our remote ancestors seems impossible to comprehend, then imagine the chasm dividing us from our descendants. This prospect includes a technologically driven Singularity, an exponentially rising curve of drastic change that promises (or threatens) to turn our kind from mortal humans into deathless posthumans. If so, can any attempt to envisage our descendants a million years hence, let alone a billion, be other than futile? Here is a vista that can seize the imagination and wring it dry, capture the heart and fill it with yearning or horror.

This book is a voyage of exploration into that barely conceivable distant future, into a time so remote that we cannot even be certain how many generations extending to the right will be required to get there. For one thing, it is not written in stone or immutable DNA that there must be four generations per century rather than one per millennium. Already, ambitious experts in gerontology—the science of aging and senescence, of our so-far-fated fall into death—are exploring ways to extend the typical life span of healthy, vigorous humans. When they succeed (for, however long it takes, it is certain that they shall succeed, since the human body is an immensely complex machine capable of repairing its own code, otherwise each baby would be born old), generations shall surely arrive more slowly, or the world would soon be choked to death with clamoring human mouths and stomachs.

So it might be that within a further century the first child will be born with the prospect of a millennium of robust life stretching before her. In this case, rather than forty thousand consecutive generations stretching to the Year Million, perhaps we shall see only one thousand or fewer. If that spoils the analogy, so much the better, since a long rich life is preferable to a short mean one, or even a short full one. Still, a mere thousand generations of our own history carries us back most of the way to the dawn of true humankind. Can we even start to conceive what the people of the Year Million might be like, supposing that our lineage does not first destroy itself or spill outward to the stars, or change into "intellects vast and cool and unsympathetic"?

This resonant and disturbing phrase is from the opening paragraph of H. G. Wells's *The War of the Worlds* (1898), in which Martians invade Victorian England. Wells was at one time, a century or so ago, one of the world's most famous writers and pundits. Before cultural arbiters decided to split radically imaginative fiction off from literature, Wells invented most of the core ideas that later formed the foundation of what we know as science fiction. But Wells was also a significant science writer, as well as a historian and an amateur in economics and political theory.

The Victorians had been obsessed for half a century with promises of new knowledge and technology, which were remaking the planet both for good and ill. Wells became rich and influential not only for his early dazzling fiction but as a great explainer, an Isaac Asimov or Carl Sagan of the turn of the twentieth century. He did not hesitate to reach back into the remote evolutionary past (Darwin's *On the Origin of Species* had established the truth of evolution only forty years earlier) or hurtle forward to the remote future. His great novel *The Time Machine* (1895) plunges a Victorian amateur scientist into the year AD 802,701, and then further still, to a dismal epoch thirty million years hence, when the rotation of our world has stalled, one face locked forever toward a vast, crimson, dying Sun scarred by dark, tumor like sunspots, then further yet, nearly to the death of Sun and Earth alike, though a dire creature cre eps on tentacles from motionless red sea to shore. That particular future, scientists have reckoned, will never happen. Within half a billion years' time, the Sun will bloat and blaze ever hotter (unless our progeny modify the very star itself, or Earth's orbit), not quite dissolving the world but blowing the atmosphere into space, evaporating oceans, and melting mountains into slag.

In November 1893, two years before *The Time Machine* astonished the world, Wells, in science-journalistic mode, cast his mind forward a thousand millennia. "The Man of the Year Million" was a choice entertainment for the *Pall Mall Gazette*, a newspaper named for novelist William Thackeray's imaginary journal that had claimed, "We address ourselves to the higher circles of society: we care not to disown it—the *Pall Mall Gazette* is written by gentlemen for gentlemen." Although Wells appears never to have reprinted that early exercise in Darwinist futurology, he self-mockingly incorporated portions, as "Of a Book Unwritten," in *Certain Personal Matters* (1901). That was the year Queen Victoria finally died after her prodigious sixty-three-year reign, the definitive end of the Victorian era, the first year of the twentieth century. The duration of her reign alone is itself sufficient to boggle the mind. Sixty-three years—how the world would be changed if all rulers remained in office so long! Victoria succeeded to the throne upon her father's death in 1837. Imagine the history of the world had her long span been the standard. Thomas Jefferson, elected president of the United States a full century before Victoria died, might then give way in 1864 to Abraham Lincoln. Were he not fated to die a year later, after the ruinous Civil War, then Lincoln would preside until his replacement in 1927, rather improbably, by Calvin Coolidge, who would see the world through the Great Depression, World War II, the Cold War, the collapse of the Soviet empire, and on into the early years of the rule of George H. W. Bush, himself likely to remain in office until the middle of the twenty-first century. None of this is strictly impossible, although my analogy is naughtily strained: Victoria was crowned at nineteen (and her granddaughter Elizabeth II at twenty-five), hardly a plausible age of succession to the chief office in a republic. Yet if this modest thought experiment with customary time is dazing, how can we possibly imagine humanity in the Year Million?

We can't. It's as simple as that. But we can try to chip at the edges of the idea of deep time, like one of those Stone Age toolmakers thirty thousand or fifty thousand years ago, mired in an eons-long annual cycle where the only change anyone remembers clearly is the familiar flux of the seasons, the bursting and then fading transformation from infancy through adulthood and then the decline to death.

Suppose we figure the passage of one year in the million-year future history of humankind (and of our successors) as a fleeting hour in a single contemporary lifetime. A million hours is just a little over 114 years. Today we have in our midst a number of supercentenarians; the oldest known human, Jeanne Calment, died at 122. Suppose humankind, as a distinct, articulate species, is sixty thousand years old. Suppose also that this species will last at least a million years from its emergence (most species persist that long before going extinct). Figuring one hour per year, our kind is now merely 7 years old, with another 107 years stretching ahead of us. We might find ourselves smiling to remember that at the age of 7, according to traditional doctrine, children reach the age of reason. Perhaps it's no accident that now we struggle with our greatest difficulties yet, and against the greatest odds, toward a world where superstition, impulsive brutality, and willful ignorance dissolve with the first stirrings of global maturity.

Wells jestingly attributes his exploration of the Year Million to a German academician, "one Professor Holzkopf," or Blockhead, fancied author of *The Necessary Characters of the Man of the Remote Future Deduced from the Existing Stream of Tendency*:

> Man, unless the order of the universe has come to an end, will undergo further modification in the future, and at last cease to be man, giving rise to some other

> type of animated being. At once the fascinating question arises, What will this being be? Let us consider for a little the plastic influences at work upon our species.

It is the key question that this book, too, will explore.

Drawing upon a rather heavy-handed Darwinism (or maybe Social Darwinism, its debased cousin), Professor Blockhead deduces certain attributes that will likely mark the man (and the woman, too, presumably) of that distant age. Freed up by machines from the urgencies and demands of labor, our offspring will grow physically feeble. Blockhead continues, "One needs wits now to live, and physical activity is a drug, a snare even; it seeks artificial outlets, and overflows in games. Athleticism takes up time and cripples a man in his competitive examinations, and in business. So is your fleshly man handicapped against his subtler brother." We laugh in disbelief at this failure to predict the narcissistic culture of the gym, of dieting, of running and swimming and skiing for health and the sheer fun of it. But is Wells entirely wrong? The athletic, Professor Blockhead asserts, "is unsuccessful in life, does not marry. The better adapted survive." Ridiculous? Is it true that many of those who preen before the mirrors, male and female alike, delay families or forego them altogether? Probably not—certainly no more so than the lazy, the harried, the much-mocked nerds whose focused mental work underpins so much of the wealth of our culture. On the other hand, Wells (or Blockhead) is trying to discern a long, long trajectory created by evolutionary pressures, the sort that shaped crabs with one enormous pincer and leopards with spots they can't change.

As a shirtsleeve intellectual, Wells foresaw the connection between intelligence, productively and profitably used, and sexual success in the post-Victorian world. If he failed to see a rise of a battalion of shy and socially incompetent nerds, perhaps that's because he was himself a nerd with a large sexual appetite and unusual success in satisfying it. In any event, Wells (or rather, Blockhead) imagined the forces of evolutionary adaptation slowly but inexorably modifying human inheritance toward the "Man of the Year Million": absorbing nutrients directly through the skin, legs withered but the manipulative hand with its sensitive fingers and thumb grown grotesquely large, the brain swollen inside its immense skull even as the Sun itself, outside their habitat domes, swells and reddens.

Incredibly, at first Wells seems to be sketching the creature popular mythology calls the Gray—the supposed alien pilot of flying saucers.

> Eyes large, lustrous, beautiful, soulful; above them, no longer separated by rugged brow ridges, is the top of the head, a glistening, hairless dome, terete [round and tapering] and beautiful; no craggy nose rises to disturb by its unmeaning shadows the symmetry of that calm face, no vestigial ears project; the mouth is a small, perfectly round aperture, toothless and gumless, jawless, unanimal, no futile emotions disturbing its roundness as it lies, like the harvest moon or the evening star, in the wide firmament of face.

But these brainiacs of the deep future are even stranger than the Grays of UFO abduction legend (which might conceivably be based on Wells's image). See them on their home ground, in the Year Million:

> There grows upon the impatient imagination a building, a dome of crystal, across the translucent surface of which flushes of the most glorious and pure prismatic colors pass and fade and change. In the center of this transparent chameleon-tinted dome is a circular white marble basin filled with some clear, mobile, amber liquid, and in this plunge and float strange beings. Are they birds?
>
> They are the descendants of man—at dinner. Watch them as they hop on their hands . . . about the pure white marble floor. Great hands they have, enormous brains, soft, liquid, soulful eyes. Their whole muscular system, their legs, their abdomens, are shriveled to nothing, a dangling, degraded pendant to their minds.

This strikes us today as an absurd, comical image of tomorrow's humanity, and probably, on reflection, it struck Wells that way, too; otherwise, why attribute this homespun evolutionary prospect to Professor Blockhead? Even so, it's hard to deny the appeal of his question, as we might rephrase it slightly today:

> *Unless our corner of the universe changes out of all recognition, our species, our stock—the kind of human person we know—will certainly undergo further modification in the future, as it has in the past, and at last cease to be human (at least as we understand the term), giving rise to some other*

> *type of consciousness. So again we face the fascinating question: What will this person be like in the Year Million and beyond, as the universe expands ever more swiftly into the darkness of space and time?*

Wells himself, once so confident of the future, offers a grim prophecy: "In the case of every other predominant animal the world has ever seen . . . the hour of its complete ascendancy has been the eve of its entire overthrow." A century or so later, that warning seems only more urgent and plausible. Anxious voices tell us that humankind is a kind of terminal pollution or plague upon the face of our blue world, as we destroy other species and the natural environment and slaughter our fellow humans by the hundred, the thousand, the million. Others, more optimistic (on happy days, I'm one of them), see in the gorgeous flowering of science in the last four hundred years proof that we are here for the long haul, extending and deepening our knowledge of the world and of ourselves. This book explores both possibilities, and more besides.

It is conceivable, now that we have learned that 95 percent of the universe remains invisible to us (its dark matter and dark energy), that consciousness will never go down under the attrition of entropy and exhaustion. Already we have hints that the universe we see is the merest fraction of its true extent, both in space and time, that the awesome ignition of our cosmos in the Big Bang might have been a bubble blown from another universe, and another before and beside that one, all the way down into the brilliance of eternity. It is not impossible that this universe is the deliberate or accidental byproduct of some colossal experiment performed in the universe preceding our own, just as scientists have conjectured means by which humans or posthumans of the future might open up entirely new universes through the creation or encoding of black holes.

Three decades after Wells's half-serious–half-facetious glimpse of the Year Million, the great British philosopher and visionary Olaf Stapledon presented an even more vertiginous cavalcade of deep time: *Last and First Men* (1930), followed seven years later by perhaps the ultimate portrait of life in the cosmos, *Star Maker*. These books are often shelved as fiction, as novels, but they have nothing in them we recognize by that description beyond the fact that their very worlds and histories are invented. Really, they are essays in projecting human understanding beyond its own limitations, from Stapledon's own frightening immediate future (the rise of Nazi Germany and the Soviet Union), through the multimillion-year sweep of human and genetically modified posthuman evolution in an increasingly reconstructed solar system, and finally to a history of those immense consciousnesses we call stars and galaxies.

The available cosmology and physics of the 1930s was hardly more developed than the sciences upon which Wells drew, yet Stapledon's reasoned dream of deep time has not yet been surpassed. It can be seen framing the essays in this book, which are themselves created by a select group of expert dreamers, chosen for their insight and knowledge, who convey to the rest of us what lies ahead of our species in the long, long voyage to the Year Million and beyond.

Is such voyaging nothing better than a disinclination to grapple with the pressing problems of today, a kind of mind candy? I don't think so. As Stapledon put it at the outset of his great journey:

> To romance of the future may seem to be indulgence in ungoverned speculation for the sake of the marvelous. Yet controlled imagination in this sphere can be a very valuable exercise for minds bewildered about the present and its potentialities. Today we should welcome, and even study, every serious attempt to envisage the future of our race; not merely in order to grasp the very diverse and often tragic possibilities that confront us, but also that we may familiarize ourselves with the certainty that many of our most cherished ideals would seem puerile to more developed minds. To romance the far future, then, is to attempt to see the human race in its cosmic setting, and to mould our hearts to entertain new values.

And if that seems rather too solemn, a touch schoolmasterly and officiously profound, then let's not forget that casting ourselves forward in controlled imagination to the deep future is a hell of a lot of *fun*.

In these fourteen specially commissioned glimpses of the future, we range from the slow convulsions and occasionally drastic lurches of the earth itself during the next million years to speculative portraits of the profoundly deep future, an era not just remoter from the Year Million than that fabled year is from the twenty-first century, but perhaps millions of times further away than we stand already from the Big Bang. We shall not restrict our gaze solely to our titular Year Million, but regard it instead as the emblem of an impossibly remote future in which humankind, or its offspring, might not only thrive but prevail, perhaps rewriting or rewiring the universe itself. This optimism is not shared by all our writers. Zoologist Dougal Dixon, who charts the vicissitudes likely to afflict our planet in the next thousand millennia, has speculated bleakly in his wonderfully illustrated and imaginative book *After Man: A Zoology of the Future:*

> Ultimately the Earth could no longer supply the raw material needed for man's agriculture, industry or medicine, and as shortage of supply caused the collapse of one structure after another, his whole complex and interlocking social and technological edifices crumbled. Man, no longer able to adapt, rushed uncontrollably to his inevitable extinction. [2]

That dire prospect remains extreme, but not impossible, as the planet's climate shifts before our gaze and threats of catastrophic global conflict continue to turn our dreams of the future into nightmares, especially with the unprecedented risks of global pandemic and new potentially lethal technologies. But apocalyptic forecasts do not speak of the inevitable, because we have choices. We are, so far, the only creatures of Earth who live forward and backward in time, in recollection and imagination. We can make plans, and while we might bring doom upon ourselves, we also hold in our clever hands and brains and passionate hearts the tools of our survival and thriving.

If our descendants do make it to the Year Million, we might hope that they share at least some of our values, some of our joy. Will they still beget children in that far-off epoch? Will they laugh at silly jokes, get drunk, stay late to work, either grumbling or driven by obsessive interest in the task at hand? Can we expect Shakespeare and Ravel to survive in their memory? Will they count in tens or twos, or leave all that to their machines? Jim Holt, a science journalist who writes frequently for the *New Yorker,* suspects that in the contest of survival between mathematics and mirth, the people of the Year Million might find a surprising winner.

All this great passage to the Year Million and beyond will be conducted under the baton of Darwin, for evolution by natural selection (including, most definitely, sexual selection) remains the great winnowing sieve, the pitiless shears that shape all life, and indeed much of culture. Physician Steven Harris, in a magisterial and exciting essay on the natural history of the future, traces some of the forces that will influence humankind's deep future.

It is often supposed by romantics that our destiny lies far from the planet Earth, an exultant and sometimes crassly imperialistic vision of a sky filled not just by remote stars but with the babbling voices of our trillion trillion children. If Earth is not destroyed soon by our own actions and the vicissitudes of nature, it really does seem plausible that our curious, exploratory nature, and the promise of riches hiding in the void, will drive and draw people into space, no matter the cost. In five essays that trace a kind of escalation of tomorrow's voyages, we look at the prospects of humanity and posthumanity in deep space through time.

Harvard-Smithsonian astrophysicist and planetary scientist Lisa Kaltenegger opens the sky to our gaze, looking at the wonderful bestiary of extrasolar worlds discovered in the last few years. Most of those have been apparently inhospitable to life and startlingly unexpected in character (blazing-hot gas giants orbiting fast around their stars, unlike the cool, slow, majestic giants of our own solar system, Jupiter and Saturn), but they are not necessarily representative specimens. The limitations of current search methods mean that these are just the worlds most easily detected. Our understanding will improve drastically, Kaltenegger explains, when space scientists of the near future cast into the vacuum first hundreds and then thousands and then tens of thousands of satellite telescopes. By the Year Million, the entire solar system will constitute a kind of colossal eye able to examine the surface of planets, earthlike and otherwise, light years away.

Mathematician, quantum theorist, and novelist Catherine Asaro is not prepared just to *look* at those worlds; she wants to go there, and suggests that perhaps relativity theory's celebrated prohibition of speeds faster than light might not be absolute after all. But even if it proves to be, elementary arithmetic makes it clear that within a million years we could, in principle, colonize the entire Milky Way galaxy. Wil McCarthy, a rocket engineer, writer, and innovative researcher in "programmable matter," wonders what life would be like for a citizen of that galaxy. Robert Bradbury, a polymath equally at home in computer programming, advanced genomics, and nanotechnology, ponders a galactic future in which entire solar systems—starting with our own—are stripped of their planets to build what he has called “Matrioshka brains.” Within these immense Sun-orbiting swarms might dwell, by the Year Million, untold trillions of human minds, no longer embodied in bone and protein, cavorting in vastly accelerated virtual realities. Perhaps, though, this radical prospect will not be the form that evolution finds most enduring. Long ago, we went out of Africa into the world; venturing into space, in economist Robin Hanson’s analysis, we might begin a wave of ferociously driven outward migration, from one star oasis to the next, to the very edge of the galaxy, perhaps leaving behind a blighted wasteland of exhausted worlds.

In such futures, what might be the place of the human mind, of consciousness? Novelist Pamela Sargent and engineer Anne Corwin consider the prospects for drastically increased, healthful longevity, a factor that will change almost everything in human experience. Technology journalist, inventor, and aerospace engineer Amara Angelica sees the Internet expanding on a cosmic scale into the Universenet. Mathematician and transrealist Rudy Rucker sports playfully in an astonishing alternative possible future, where machines have been discarded, replaced by life and mind seeping into everything around us, down to the level of atoms and perhaps deeper. It is a vision so audacious that it seems like a wild joke. But how would our Pleistocene ancestors have regarded a prophetic glimpse of people gazing into a flat screen bright with images of distant places or strange squiggles, hammering with their fingers at a notched board, or speaking into the air as the squiggles dance responsively?

Finally, taking our exploration far beyond the Year Million, to the ends of the expanding universe, across the river of time and into the branches of an endlessly exfoliating multiverse, astrophysicist and novelist Gregory Benford, cosmologist Sean Carroll, and visionary writer George Zebrowski peer into the very deepest future. What they find might seem disheartening—a cosmos evaporating and corroding even as its parts fling themselves so far asunder that they can no longer detect each other. Yet this aching, distant void need not be the end of the story of creation, of the narrative of humankind, and our children. Perhaps a trillion years beyond the Year Million, or sooner, we will be building new universes, finding our way across the boundaries of spacetime into realms of which, today, we can only dream and yearn.

The Expanding Human Universe

Chapter 1

The Laughter of Copernicus

Jim Holt

In the Year Million, most of the things we're familiar with today will have disappeared. But some will survive. We can be pretty confident that among those will be numbers and laughter. That is good because, in their different ways, numbers and laughter make life worth living. So it will be interesting to ponder their status in the Year Million. But before we get to that, let me tell you why I am so sure they will still be around, when almost everything else we know today will have either disappeared or evolved into something unrecognizable.

In general, things that have been around for a long time are likely to remain even longer. Conversely, things of recent origin probably won't. Both of these conclusions flow from the Copernican principle, which says, in essence, *you're not special.* If there's nothing special about our perspective, then we're unlikely to observe any given thing at the very beginning or end of its existence. Let's say you go to see a Broadway play. No one can be sure exactly how long the play will run. It could close after a few nights or continue for many years. But you do know that, of all the people who will see it, 95 percent of them will be among neither the first 2.5 percent nor the last 2.5 percent to do so. Therefore, if you're not "special"—that is, if you're just a random member of the play's total audience—then you can be 95 percent sure that you don't fall into either of these two tails of the probability distribution.

If the play has already had *n* performances at the point in its run when you happen to see it, then you can be 95 percent sure that it has no more than 39 times *n* performances to go and no fewer than *n* divided by thirty-nine. (This is a matter of elementary arithmetic: the upper limit keeps you out of the first 2.5 percent of the total audience, and the lower limit keeps you out of the last 2.5 percent.) With nothing more than the Copernican principle and a grade-school calculation, you can come up with a 95 percent confidence interval for the longevity of something like a Broadway play.

That is fairly amazing.

Richard Gott III, an astrophysicist at Princeton University, is the one who pioneered this sort of reasoning. In a 1993 paper published in *Nature,* Gott calculates the expected longevity of our species.[1] Humans have been around for some 200,000 years. So if there is nothing special about the moment at which we observe our species, then we can be 95 percent sure that it will persist for at least 5,100 years ($\frac{1}{39}$ times 200,000) but will disappear within 7.8 million years (39 times 200,000). This calculation, Gott notes, gives *Homo sapiens* an expected total longevity comparable to other hominid species (*Homo erectus,* our ancestor species, lasted 1.6 million years) and to mammal species in general (whose average span is 2 million years). It also gives us a decent shot at being around in the Year Million.

But what else will be around then? Consider something of recent origin, like the Internet. The Internet has existed for about twenty-five years now (as I learned by going onto the Internet and looking at Wikipedia). That means, by Copernican reasoning, we can be 95 percent certain that it will continue to be around for another 7-plus months but that it will disappear within 975 years. So, in the Year Million, there will probably be nothing recognizable as the Internet. Ditto for baseball, which has been around for a little more than 2 centuries now. Ditto for what we call "industrial technology," which, having come into existence a few hundred years ago, is likely to be superseded by something strange and new in the next 10,000 years. Nor, by the same Copernican reasoning, is organized religion a good bet to survive to the Year Million or anytime near it.

To find something that will pretty certainly still be around in the Year Million, we are obliged, paradoxically enough, to go back much further in our natural history. Again, this is because, as Gott puts it, "Things that have been around for a long time tend to stay around for a long time."[2] And if we could cast a look back several million years into the past, we would see, among other things, laughter and numbers. How do we know this? Because we share both laughter and a sense of number with other species today, and therefore with common ancestors that existed millions of years ago.

Take laughter. Chimpanzees laugh. As Charles Darwin notes in "The Expression of Emotions in Man and Animals" (1872), "If a young chimpanzee be tickled—the armpits are particularly sensitive to tickling, as in the case of our children—a more decided chuckling or laughing sound is uttered; though the laughter is sometimes noiseless." Actually, what primatologists call chimp laughter is more like a breathy pant, induced not only by tickling but also by rough-and-tumble play, games of chasing, and mock attacks—just as with children prior to the emergence of verbal joking at age five or six.[3] But does primate humor ever rise above sheer physicality? The researcher Roger Fouts reports that Washoe, a chimp who was taught sign language, once urinated on him while riding on his shoulders, signing "funny" and snorting but not laughing.

The human and chimpanzee lineages split off from each other between five and seven million years ago.[4] On the reasonable assumption that chimp and human laughter are homologous rather than having evolved independently, we can figure that laughter must be at least five to seven million years old. (It is probably much older: orangutans also laugh, and their lineage diverged from ours about fourteen million years ago.) So, by the Copernican principle, laughter is quite likely to be around in the Year Million.

Now take number. Chimps can also do elementary arithmetic, and they have even been trained to use symbols like numerals to reason about quantity. And a sense of number is not confined to primates. Researchers have found that animals as diverse as salamanders, pigeons, racoons, dolphins, and parrots demonstrate the ability to perceive and represent number. A few years ago, researchers at M.I.T. found that macaque monkeys had specialized "number neurons" in the brain region corresponding to the location of the human number module. Evidently, the number sense has an even longer evolutionary history than that of laughter.[5] So again, by the Copernican principle, we can be quite certain that numbers will be around in the Year Million.

Of the cultural wonders of our world, numbers and laughter are two of the oldest. As such, they are likely to survive the longest, quite probably beyond the Year Million. One might draw an analogy to the Seven Wonders of the Ancient World. When this list was drawn up (the earliest extant version dates to about 140 BCE), the oldest wonder on it by far was the pyramid of Giza, which went back to about 2500 BCE. The other six wonders—the Hanging Gardens of Babylon, the temple of Artemis at Ephesus, the statue of Zeus at Olympia, the mausoleum at Halicarnassus, the Colossus of Rhodes, and the lighthouse at Alexandria—were almost two millennia newer. And which of the Seven Wonders still survives today? The pyramid of Giza. All the rest have disappeared, done in by fires or earthquakes.

Laughter and number are like the pyramid in their expected longevity. And a good thing, too, because they lie at the core, respectively, of humor and mathematics, and these make life bearable for the nobler spirits among us. Bertrand Russell recounts in his autobiography that, as an unhappy adolescent, he frequently contemplated suicide.[6] But he did not go through with it, he tells us, "because I wished to know more of mathematics." Woody Allen's character in the film *Hannah and Her Sisters* is similarly given to suicidal thoughts, but he is pulled back from the brink when he goes to a revival cinema and sees the Marx Brothers in *Duck Soup* playing on the helmets of the soldiers of Freedonia like a xylophone. If our descendants in the Year Million are to find existence worth the bother, then they had better retain laughter and mathematics.

But what will their mathematics look like? And what will make them laugh?

The first question might seem the easier to answer. Mathematics, after all, is supposed to be the most universal part of human civilization. All terrestrial cultures count, so all terrestrial cultures have number. If intelligent life exists elsewhere in the cosmos, then we would expect the same to be true for them. The one earmark of civilization likely to be recognized across the universe is number. In Carl Sagan's novel *Contact*, aliens in the vicinity of the star Vega beam a series of prime numbers toward Earth. The book's heroine, who works for SETI (Search for Extraterrestrial Intelligence), realizes with a frisson that the prime-number pulses reaching her radio telescope must have been generated by some form of intelligent life. But what if the aliens beamed their *jokes* at us, instead? Probably we wouldn't be able to distinguish the jokes from the background noise.

Indeed, we can barely distinguish the jokes in Shakespeare's plays from the background noise. (Seriously, have you ever laughed out loud during a Shakespeare play? His audiences used to roar.) Just as nothing is more timeless than number, the core of mathematics, nothing is more parochial and ephemeral than humor, the core of laughter. So, at least, we imagine. We are quite confident that a civilization a million years more advanced than our own would find our concept of number intelligible, and we theirs. But their jokes would have us scratching our heads in puzzlement, and vice versa if they still retain heads. As for all the bits of culture in between—ranging from literature and art (at the more parochial end) to philosophy and physics (at the more universal end)—sure, they might be recorded and available, perhaps on the Universenet [see Rudy Rucker's and Amara Angelica's essays—Ed.] but will anyone bother to access them, or understand them if they do?

That's how we see matters at the moment. In the Year Million, though, I think the perspective will be precisely the reverse. Humor will be esteemed as the most universal aspect of culture. And number will have lost its transcendental reputation and be looked upon as a local artifact, like a transient computer operating system or an accounting scheme. If I am right, then, SETI scientists should not be listening for prime numbers or the digits of pi, but for something quite different.

Let's return to number for a moment. (Those who prefer jokes should feel free to skip to the end.) In 1907, when he was in his thirties, Bertrand Russell penned a gushing tribute to the glories of mathematics. "Rightly viewed," Russell wrote, mathematics "possesses not only truth, but supreme beauty—a beauty cold and austere, like that of sculpture, without appeal to any part of our weaker nature, without the gorgeous trappings of painting or music, yet sublimely pure, and capable of a stern perfection such as only the greatest art can show." [7] These lines, which play up the transcendent image of mathematics, are often quoted in mathematical popularizations. What one seldom encounters in such books, however, is the rather different view that Russell expressed in his late eighties, when he dismissed his youthful rhapsodizing as "largely nonsense."

Mathematics, the aged Russell writes, "has ceased to seem to me non-human in its subject-matter. I have come to believe, though very reluctantly, that it consists of tautologies.

I fear that, to a mind of sufficient intellectual power, the whole of mathematics would appear trivial, as trivial as the statement that a four-footed animal is an animal."[8] So, in the course of his life, Russell underwent an evolution in his thinking about mathematics. I think our civilization will have undergone a similar evolution by the Year Million. (Yes, Virginia, phylogeny sometimes recapitulates ontogeny.) Our descendants will view mathematics merely as an elaborate network of tautologies, of strictly local import, that has proved expeditious as a bookkeeping scheme for coping with the world.

If mathematics is essentially trivial, then its triviality ought to be most apparent at the elementary level, before the smoke and mirrors of higher theory have had a chance to do their work. So let's focus on the elementary. The most fundamental objects in mathematics, everyone would agree, are the counting numbers: 1, 2, 3, etc. Among such numbers, the prime numbers—those beamed by the aliens in *Contact*—are supposed to be special. A prime is a number that cannot be split up into smaller factors. (Another way of putting this is to say that a prime is a number that can be evenly divided only by itself and the number one.) The first few primes are 2, 3, 5, 7, 11, 13, 17, 19, 23, 29, 31, 37. . . . The primes are the atoms of arithmetic: all the rest of the numbers, called composite numbers, can be built up by multiplying primes together in various combinations. For example, the number 666 can be obtained by multiplying 2 times 3 times 3 times 37. One can prove, with just a little trouble, that every composite number can be put together in one and only one way as a product of primes. This is often called the "fundamental theorem of arithmetic."

So far so good: everything looks tautological enough. Let's move to the next obvious question: How many prime numbers are there? This question was posed by Euclid in the third century BCE, and the answer is contained in proposition twenty of his *Elements*: there are infinitely many primes. Euclid's proof of this proposition is perhaps the first truly elegant bit of reasoning in the history of mathematics. It can fit into a single sentence: if there were only finitely many primes, then, by multiplying them all together and adding 1, you would get a number that could not be divided by any prime at all, making it an additional prime, which is impossible because we'd already have used up all the primes.

Once we know that the primes go on forever, the next question that naturally arises is: How are these atoms of arithmetic scattered among the rest of the numbers? Is there a pattern? Primes turn up rather frequently among the smaller numbers, but they get scarcer as you move through the number sequence. Of the first ten numbers, four are prime. Of the first hundred numbers, twenty-five are prime. Jumping on a bit, of the hundred numbers between 9,999,900 and 10,000,000, nine are prime; among the next hundred numbers, from 10,000,000 to 10,000,100, only two are prime (10,000,019 and 10,000,079). It is possible to find stretches of numbers as long as you please that are completely prime-free. But there are also very large primes that clump together, like 1,000,000,009,649 and 1,000,000,009,651. (Primes differing by only two in this way are called "twin primes"; it is an open question whether there are infinitely many of them.) Prime numbers seem to crop up almost at random, sprouting like weeds among the rest of the numbers. "There is no apparent reason why one number is a prime and another not," declared Don Zagier, one of the greatest living mathematicians, in his inaugural lecture at Bonn University in 1975. "To the contrary, upon looking at these numbers one has the feeling of being in the presence of one of the inexplicable secrets of creation."[9]

Despite their simple definition, the primes appear to have a complex and timeless reality all their own, one quite independent of our minds. They are transcendently mysterious in a way that the proposition "a four-footed animal is an animal" is not. But are they completely lawless? That would be very surprising, given their role as the building blocks of arithmetic. And in fact they do obey a law. But to find this law, strangely, one must ascend many stories in the edifice of mathematics: from the humble counting numbers through the integers, the fractions, and the real numbers all the way to complex numbers with "imaginary" parts. (Historically, that ascent took longer than two millennia.) And then, at that lofty level, one runs into a conundrum known as the Riemann zeta hypothesis.

By the near-unanimous judgment of mathematicians, the Riemann zeta hypothesis is the greatest unsolved problem in mathematics. It may be the most difficult problem even conceived by the human mind. The Riemann in question is Bernhard Riemann, a nineteenth-century German mathematician. "Zeta" refers to the zeta function, a creature of higher mathematics that, he realized, held the secret of the primes. In 1859, in a brief but exceedingly profound paper, Riemann put forward a hypothesis about the zeta function. If his hypothesis is true, then there is a hidden harmony to the primes, one that is rather beautiful. If it is false, then the music of the primes could turn out to be somewhat ugly, like that produced by an orchestra out of balance.

Which will it be? For the last century and a half, mathematicians have been striving in vain to resolve the Riemann zeta hypothesis. In a celebrated speech delivered in 1900 before an international conference in Paris, mathematician David Hilbert included it on his list of the twenty-three most important problems in mathematics. He later declared it to be the most important "not only in mathematics, but absolutely the most important."[10] The Riemann hypothesis was the only problem on Hilbert's list to survive last century undispatched. In 2000, on the one hundredth anniversary of Hilbert's speech, a small group of the world's leading mathematicians held a press conference at the Collège de France to announce a fresh set of seven "Millennium Problems," the solution of any of which would be rewarded with a prize of one million dollars, courtesy of the Clay Institute for Mathematics, founded by the Boston investor Landon T. Clay. To no one's surprise, the Riemann hypothesis made this list too.

The Riemann zeta hypothesis is more than just the key to understanding the primes. So central is it to mathematical progress that its truth has simply been assumed—perhaps rashly—in the proofs of thousands of theorems (which are said to be "conditioned" on the hypothesis). If it turns out to be false, then the part of higher mathematics that is built upon it will collapse. (Fermat's famous last theorem, finally proved in 1995, played no such role as a structural component of mathematics and hence was far less important.)

The zeta function, fittingly, has its origins in music. If you pluck a violin string, it vibrates to create not only the note to which it is tuned but also all possible overtones. Mathematically, this combination of sounds corresponds to the infinite sum $1 + \frac{1}{2} + \frac{1}{3} + \frac{1}{4} + \ldots$, which is known as the "harmonic series." If you take every term in this series and raise it to the variable power *s*, you get the zeta function:

$$\text{zeta}(s) = \left(\frac{1}{2}\right)^s + \left(\frac{1}{3}\right)^s + \left(\frac{1}{4}\right)^s + \ldots$$

This function was introduced around 1740 by Leonhard Euler, who proceeded to make a striking discovery. He found that the zeta function, an infinite *sum* running though all the numbers, could be rewritten as an infinite *product* running though just the primes:

$$\text{zeta}(s) = \frac{1}{(1-\frac{1}{2}^s)} \times \frac{1}{(1-\frac{1}{3}^s)} \times \frac{1}{(1-\frac{1}{5}^s)} \times \frac{1}{(1-\frac{1}{7}^s)} \times \frac{1}{(1-\frac{1}{11}^s)} \times \ldots$$

Although he was the greatest mathematician of his time, Euler did not fully grasp the potential of his infinite product formula. "Mathematicians have tried in vain to this day to discover some order in the sequence of prime numbers," he wrote, "and we have reason to believe that it is a mystery into which the human mind will never penetrate."[11]

A half century later, Carl Friedrich Gauss made the first real breakthrough since Euclid in understanding the prime numbers. As a boy, Gauss enjoyed tallying up how many primes there were in each block of a thousand numbers. Such computations were a good way to beguile "an idle quarter of an hour," he wrote to a friend, "but at last I gave it up without quite getting through a million."[12] In 1792, when he was fifteen years old, Gauss noticed something interesting. Although the primes cropped up seemingly at random, there did appear to be some regularity to their overall flow. A good estimate for how many primes up to a given number could be obtained by dividing that number by its natural logarithm. Suppose, for example, you want to know how many primes there are up to a million. Take out your pocket calculator, punch in 1,000,000 and divide it by the ln(1,000,000). Out pops 72,382. The actual number of primes up to a million is 78,498, so the estimate is off by about 8 percent. But the percentage error heads toward zero as the numbers get bigger.

What Gauss had discovered was "the coin that Nature had tossed to choose the primes" (in the words of the British mathematician Marcus du Sautoy). It was uncanny that this coin should be weighted by the natural logarithm, which arose in the continuous world of the calculus and would seem wholly unrelated to the chunky world of the counting numbers. Gauss could not prove that the log function would continue to describe the waning of the primes through the entire infinity of numbers; he was just making an empirical guess. Nor could he explain its inexactness: why it failed to say precisely where the next prime would turn up.

It was Riemann who made the deep connections necessary to dispel the lingering illusion of randomness. In 1859, in a paper of fewer than ten pages, he made a series of moves that cracked the mystery of the primes. He began with the zeta function, venturing beyond Euler, who had seen this function as ranging only over "real" values. (The real numbers, which correspond to the points on a line, comprise the whole numbers, both positive and negative; rational numbers, which can be represented by fractions; and irrational numbers, like pi or *e*, which can be represented by nonrepeating decimals.) Riemann enlarged the zeta function to take in the complex numbers, those having both a real and an "imaginary" part involving *i*, the square root of –1.

Since the complex numbers are two-dimensional, they can be graphed as a plane. In extending the zeta function over this complex plane, Riemann in effect created a vast imaginary landscape—the zeta landscape—consisting of mountains, hills, and valleys that stretched forever in every direction. The most interesting points in the zeta landscape, he found, were the ones with zero altitude—that is, the sea-level points. These points are called the "zeros" of the zeta function, since they correspond to those complex numbers that, if plugged into the zeta function, yield the output zero. Using these complex zeros, of which there are infinitely many in the zeta landscape, Riemann was able to do a marvelous thing: he produced, for the first time ever, a formula that described *exactly* how the infinity of primes arranged themselves in the number sequence.

This discovery opened up a metaphorical dialogue between mathematics and music. Before Riemann, only random noise could be heard in the primes. Now there was a way to listen to their music. Each zero of the zeta function, when plugged into Riemann's prime formula, produces a wave resembling a pure musical tone. When these pure tones are all combined, they reproduce the harmonic structure of the prime numbers. Riemann found that the location of a given zero in the zeta landscape determines the pitch and volume of its corresponding musical note. The farther north the zero is, the higher the pitch. And—more important—the farther east it is, the greater the loudness. Only if all the zeros lie in a fairly narrow longitudinal strip of the zeta landscape will the orchestra of the primes be in balance, with no instrument drowning out the others. But Riemann went further. After navigating just a tiny part of the infinite zeta landscape, he boldly asserted that all of its zeros were precisely arrayed along a "critical line" running from south to north. And this is the claim that subsequently became known as the Riemann zeta hypothesis.

"If Riemann's Hypothesis is true," writes du Sautoy, "it will explain why we see no strong patterns in the primes. A pattern would correspond to one instrument playing louder than the others. It is as if each instrument plays its own pattern, but by combining together so perfectly, the patterns cancel themselves out, leaving just the formless ebb and flow of the primes."[13] There is something magical in the way the infinity of zeros in the zeta landscape collectively controls how the infinity of primes occur among the counting numbers: the more regimented the zeros are on one side of the looking glass, the more random the primes appear on the other.

But are the zeros as perfectly regimented as Riemann believed? If the Riemann hypothesis is false, a single zero off the critical line would suffice to refute it. Calculating where these zeros lie is by no means a trivial matter. Riemann's own navigation of the zeta landscape revealed that the first few sea-level points lined up the way he expected they would. In the early twentieth century, hundreds more zeros were computed by hand. Since then, computers have located billions of zeros, and every one of them lies precisely on the critical line. One might think that the failure to find a counterexample to the Riemann hypothesis so far increases the odds that it is true. This is a matter of some controversy. There are, after all, infinitely many zeta zeros, and it may be that they only reveal their true colors in unimaginably distant reaches of the zeta landscape—reaches whose exploration might lie well beyond the Year Million. Those who blithely assume the truth of the Riemann conjecture should keep in mind an interesting pattern in the history of mathematics: whereas long-standing conjectures in algebra (like Fermat's theorem) typically turn out to be true, long-standing conjectures in analysis (like the Riemann conjecture) often turn out to be false.

The majority of mathematicians today who cleave to the Riemann hypothesis do so primarily on aesthetic grounds: it is simpler and more beautiful than its negation, and it leads to the most "natural" distribution of primes. "If there *are* lots of zeros off the line—and there might be—the whole picture is just horrible, horrible, very ugly," the mathematician Steve Gonek has said.[14] The hypothesis is unlikely to have any practical consequences, but that is of little import to the mathematicians who pursue it. "I have never done anything 'useful,'" G. H. Hardy brags in *A Mathematician's Apology*. "No discovery of mine has made, or is likely to make, directly or indirectly, for good or ill, the least difference to the amenity of the world."[15] (As it turned out, he was wrong about that.)

Mathematicians like Hardy acknowledged two motives. One is the sheer pleasure of doing mathematics. The other is the sense that mathematicians are like astronomers peering out at a Platonic cosmos of numbers, a cosmos that transcends human culture and any other civilizations there might be, either now or in the future. Hardy adds, "317 is a prime, not because we think so, or because our minds are shaped in one way rather than another, but *because it is so*, because mathematical reality is built that way." Alain Connes, a French mathematician who is widely deemed to be a leading candidate to prove the Riemann hypothesis, is also unabashed in his Platonism. "For me," Connes has said, "the sequence of prime numbers . . . has a reality that is far more permanent than the physical reality surrounding us."[16]

But will this remain true in the Year Million? Prime numbers will probably retain this transhuman reputation only until we come to understand them more completely. Then we will see that, like the rest of mathematics (or like religion, for that matter), they are man-made, a terrestrial artifact. So when can we expect the great devalorization? Paul Erdos, the late gypsy genius of mathematics, is reputed to have declared, "It will be another million years, at least, before we understand the primes." The Copernican principle yields a rather different estimate. The Riemann zeta conjecture has been open since it was first posed by Riemann himself, 148 years ago. That means we can be 95 percent certain that it will survive as an open problem for another four years or so ($\frac{1}{39} \times 148$), but that it will be dispatched within the next six millennia (39 × 48)—well short of the Year Million. If and when it occurs, the prime numbers will be finally stripped of their cosmic otherness.

The prime numbers define the zeta function; the zeta function determines the zeros; and the zeta zeros contain the secrets of the primes. Resolving the Riemann zeta hypothesis will close this tight little circle completely, rendering the mystery of the primes as tautological as the statement that a four-footed animal is an animal. My prediction, then, is that long before the Year Million, mathematicians will awaken from their collective Platonist dream. No one will give a thought to beaming prime numbers throughout the cosmos. Our descendants will dismiss them in the manner of the hero of Bertrand Russell's story, "The Mathematician's Nightmare," by saying, "Avaunt! You are only Symbolic Conveniences!"[17]

And how about the future of laughter? As I observed earlier, nothing is deemed to be more parochial and ephemeral than humor. Or more lowly. For much of human history, the comical has been a mix of lewdness, aggression, and mockery. As for the peculiar panting and chest-heaving behavior to which it gives rise, it has traditionally been viewed as a "luxury reflex" serving no obvious purpose.

In recent years, though, practitioners of the fanciful art of evolutionary psychology have been more inventive, coming up with Darwinian rationales for laughter. One of the most plausible comes from the neuroscientist V. S. Ramachandran, who advanced what might be called the "false-alarm" theory of laughter. A seemingly threatening situation presents itself; you go into the fight-or-flight mode; the threat proves spurious; you alert your (genetically close-knit) social group to the absence of actual danger by emitting a stereotyped vocalization—one that is amplified as it passes, contagiously, from member to member.[18]

Once this mechanism had developed through evolution, it could be hijacked for other purposes—expressing hostility toward (and superiority vis-a-vis) other social groups or ventilating forbidden sexual impulses within one's own. But at the core of the original false-alarm mechanism of laughter lies *incongruity:* the incongruity of a grave threat revealing itself to be trivial; of *something* evaporating into *nothing.* And in the evolution of humor over the millennia, the perception of incongruity has played a more and more dominant role. Laughter, at its highest, is now regarded as an intellectual emotion. Indeed, the upper bound of the evolution of jocularity might be the Jewish joke, where a Talmudic playfulness toward language and logic reigns. (Think of your favorite Groucho Marx or Woody Allen line.) By this intellectualist view, the greatest stimulus to laughter is pure, abstract incongruity. Every good joke, as Schopenhauer held, is a disrupted syllogism. (A classic example: "The important thing is sincerity. If you can fake that, you've got it made.") And incongruity is the opposite of boring old tautology. And just as universal.

That is why I think humor and mathematics will have changed places by the Year Million. But what will jokes look like in that distant day? The higher laughter is called forth when we see an incongruity resolved in some clever way, resulting in an emotional shudder of pleased recognition. We imagined we were apprehending something odd and incomprehensible, but suddenly we find ourselves holding nothing. The Riemann zeta hypothesis, when it is finally dispatched in the aeons to come, will provide just such a resolution. Amid peals of laughter, the Platonic otherness of the primes will dissolve into trivial tautology. It is sobering to think that what is today regarded as the hardest problem ever conceived by the human mind might well prove to be, in the Year Million, a somewhat broad joke, fit for schoolchildren.

Chapter 2

A Changing Earth

Dougal Dixon

One thing we know for certain about the Year Million: the world won't much resemble the way it is today. That doesn't mean the adapted creatures of that distant time must be entirely alien to our experience of life on Earth. (Unless our descendants are the ones responsible for deliberately changing them, as seems quite possible.)

Whether we like it or not, the planet's climate is already changing, visibly, and at a remarkable rate. The reasons for this change seem apparent. It's mostly an unintended side effect of human interventions in natural processes. The increase of greenhouse gases, mostly carbon dioxide and methane, is well documented and beyond dispute. For tens or hundreds of millions of years—back in the Paleozoic and Mesozoic eras—the carbon dioxide component of the Earth's atmosphere was absorbed by living things. Gradually it stabilized into deposits of coal and oil, buried beneath the Earth's surface. With our technology and industry, we have been burning these fuel reserves for more than a century, and in mere decades all this carbon dioxide that took so long to capture has been returned to the atmosphere. Additionally, left to itself the ocean and its plankton could absorb great quantities of increased carbon dioxide, but pollution of the sea is disturbing the ability of the oceans and their life-forms to keep the atmospheric composition stable. Little wonder, then, that the composition of the atmosphere, and hence its physical properties and the climatic conditions it supports, is changing.

We have become fatigued by this news, but the latest predictions still make alarming reading. Generally increasing temperatures will force climate zones farther toward the poles, at a possible rate of six kilometers per decade; within this greater trend, a number of distinct patterns emerge. Temperature and rainfall predictions for the next century show dramatic changes across the whole of Africa, tropical South America, the Middle East and central Asia, and the northernmost reaches of North America. Such places as Brazil, West Africa, India, and Southeast Asia—an estimated area of from 12 to 39 percent of the land surface—will develop extreme climates, that is, novel combinations of temperature, pressure, and moisture that currently exist nowhere else in the world.

Along the Andes and Central America, central Africa, southern Australia, and the far north of Asia—an estimated 10 to 48 percent of land area—climate zones will just cease to exist. Climatic change usually implies the slow shift northwards or southwards of zones of climate, so slow that the animal life adapted to them might be able to follow them to the new regions. Now, by contrast, we are talking about a complete change of specific combinations of factors—humidity, temperature, etc.—that make up a particular climate. This prospect is particularly disturbing because animal life in these regions will have nowhere to which they can migrate. As a result, between 15 and 37 percent of all species—maybe a million species in all—will be tipped into extinction. If anybody wants to observe a mass extinction in action, this is it! A geologist of the Year Million would see this disappearance as an abrupt break in the fossil record, rivaling that of the end of the Permian or the end of the Cretaceous.

What makes these predictions so appalling is that these changes will happen before our eyes, within the span of a human lifetime. The longer-term effect—toward the Year Million—will be much greater, and far less predictable.

History

The geological history of our planet is full of such climate changes, and even without human intervention that pattern is bound to continue into the far future. During the Carboniferous period 300 million years ago, Earth's atmosphere consisted of something like 35 percent oxygen—much richer than today's 23 percent—which gave rise to gigantic arthropods, spiderlike creatures able to cope with this oxygen-rich breathing mixture. The most significant period of global warming was about 55 million years ago, at the boundary between the Paleocene and the Eocene. Sea temperatures rose by between 5 and 8 degrees Celsius over a few thousand years, and Arctic waters became almost tropical. This climate shift resulted in the extinction of important oceanic foraminifera and a huge turnover of terrestrial mammal fauna. An increase in carbon dioxide in the atmosphere is blamed for this event, but the mechanisms that caused it are obscure. Then, about three million years ago, the closing of the gap between North and South America disrupted the existing oceanic circulation, which, combined with the global movements of continents and irregularities in the orbit of the Earth, triggered the Pleistocene Ice Age.

In the future, once the atmosphere has settled down after our current tinkering, and a new equilibrium has been established, the effects of the natural forces that have determined climate changes throughout deep time will reappear. Within the next million years, such factors might well lead us into another ice age, despite whatever we have done to increase the general temperature. On the other hand, dramatic increases in greenhouse gases caused by natural events like volcanic eruptions or the release of methane from oceanic deposits might accelerate the process of global warming and prevent an ice age.

Which direction the changes will go in is impossible to predict.

Tectonic Change

Earth's surface is on the move. This has always been the case, even though the idea has taken hold only in the last half century.

In the mid-twentieth century, an old heretical concept—continental drift—was combined with a new scientific discovery—seafloor spreading—to produce the all-embracing concept of plate tectonics. Earth's surface consists of plates like the panels of a football, and their sluggish movements carry the continents to and fro. What resulted from the past million years of plate shifts, and what will happen by the Year Million? Oddly, the answer is: not much.

In 1492, when Christopher Columbus crossed the Atlantic Ocean and became the first recorded European to set foot in North America, the trip took him seventy days. Repeating the same journey today would take him a little longer—a very little longer. For the Atlantic Ocean is ten meters wider (about thirty-three feet) now than it was five hundred years ago. Simple mathematics shows how much movement will take place over the next million-year period: twenty kilometers, or twelve miles. That would be no big deal on the map of the world. Certainly it would be meager compared with the changes to world geography that have taken place since the age of the dinosaurs.

The fastest known plate movement is that of the Indo-Australian plate, heading northwards. There have been various speed estimates, none faster than ten centimeters (about four inches) per year. At this rate, after a million years Australia will have moved one hundred kilometers (about sixty-two miles) north toward Asia. This distance is not great in the grand scheme of things, and by itself would not move the Australian continent from one climate zone to another. Near-term climate changes from other causes will have much more influence than continental drift.

If, however, we look at the extreme long term—way beyond the Year Million—certainly we shall see great changes. Australia will impact fully with the Asian continent in about fifty million years. The on-off attachment between North and South America through Central America and the Isthmus of Panama will continue to open and close. North and South America are currently connected by a very narrow neck of land. The mountains that form this neck are continually rising and eroding away, and for the last few million years the neck has alternated between being a complete land bridge, a string of islands, and a swathe of open ocean. While open, it will facilitate an equatorial current between the Atlantic and Pacific oceans, with all sorts of implications for world climates. Looking even further into the future, we would see the moving continents coming together again to form another Pangaea, merging all of the planet's landmasses into a single supercontinent. But that will occur well beyond the age of human civilization, however we calculate it.

The Mediterranean region is particularly tectonically active. A look at the map of Europe shows the result. Over the last few hundreds of millions of years, the Tethys Ocean that once existed between Africa and Europe disappeared as Africa moved northwards. That continent eventually crashed into the southern edge of Europe, squeezing out the great ocean, leaving only the Mediterranean and the drying puddles of the Black, Caspian, and Aral seas, and then began shearing away to the east.

The initial impact thrust up new mountain ranges—the Atlas Mountains in Morocco, the Apennines in Italy, the Alps and the Carpathians on mainland Europe—and then twisted them into great loops by the shearing action. The eastern part of Europe was pushed away eastwards, while the western part remained relatively stationary, resulting in great strains being put on the continent. This stretching opened north-south rift valleys across Europe—the Rhône Valley, the Rhine Valley, and the graben structure (a kind of collapsed trench) beneath the North Sea, where oil deposits have been found. The Iberian Peninsula—Spain and Portugal—came ripping out of the side of France and swung around southward, opening the Bay of Biscay between them and crumpling up a stretch of land to form the Pyrenees along the hinge. Two vast islands—Corsica and Sardinia—were plucked from the south coast of France and pulled southward into the sea.

Just look at a physical map of Europe and all of this motion will spring out as obvious. The frequency of earthquakes in the area, the large number of active volcanoes, and the subduction zone and island arc located off Greece show that these continental movements are still occurring. In the next million years this activity will continue, wrenching the continent of Europe still farther apart, throwing up the boundary mountains to new heights, and narrowing the Mediterranean even more.

Of course this dramatic language—"ripping," "wrenching," "twisting"—disguises the fact that these changes occur fantastically slowly by human standards. We are talking about centimeters per year. Even so, the action of plate movement can be noticeable, and even devastating, in the short term. If the movement of a particular plate is stalled for a long time, tension builds up and, when finally released, lets rip in destructive earthquakes along the border with the adjacent plate. Likewise, volcanic eruptions are produced as a side effect of this activity, and they can have a terrifying local effect on human population.

A more widespread effect is possible. Lava "trapps" (a trapp is a massive plateau built up from lava flows over a long period of time) can erupt continuously over thousands of years, as they did in India at the end of the Cretaceous period and in Siberia at the end of the Triassic. In a great outpouring, particularly runny basaltic lava flows long distances and builds up plateaus several kilometers thick, burying whole sections of continent. Basaltic lava usually erupts at constructive plate margins, like those along oceanic ridges. This kind of lava forms in Iceland, where, luckily, eruptions occur mostly at the bottom of the sea. These major historical eruptions took place at times that we regard as the ending of one geological period and the beginning of another, marked by mass extinctions and a turnover of animal life, suggesting that they had some widespread influence on the Earth as a whole. Certainly the change in atmospheric composition caused by such intense eruptions would have had a marked effect on climates worldwide.

These events are infrequent and, given the present state of geological knowledge, unpredictable. Can it happen within the next million years? So far, science has no way of knowing in advance. Such a catastrophe would bring widespread disaster and alarm; the human species has a sorry tendency to be ill prepared for anything untoward. Although ours is the only species to have evolved the ability to contemplate its own death, we do strangely little to anticipate the events that may bring it about.

Magnetic Change

The structure and the internal movements of the Earth have another subtle effect. Earth's core consists of two parts—a dense, solid inner part and an outer liquid portion—rotating at different speeds. One product of this rotation is the Earth's magnetic field. The moving liquid iron of the core cuts through the existing magnetic field, producing electric currents that generate further magnetic fields. (A range of mechanisms is at work here, the motion of the core being the greatest of them.) This magnetism spreads its influence throughout the globe and manifests itself on the surface where, as a handy by-product, it allows navigators to detect the direction of North. Earth's magnetic field is not a pure dipole with distinct north and south poles, the kind we find in a fridge magnet. It is much more complex. Still, the dipole component is stronger than any other magnetic pattern detectable across the surface of our planet, which allows us those useful magnetic compasses.

Nor is Earth's magnetic field constant; in fact, it is forever changing. The actual position of the North Magnetic Pole swings about in the Arctic Ocean, moving about fifteen kilometers (about nine miles) per year. Mapmakers estimate the movement of the North Magnetic Pole and publish calculations of its rate of movement.

More dramatic still, once in a while the global magnetic field reverses itself. The North Pole becomes a south magnetic pole and vice versa. Proof came from the bottom of the ocean, revealed during the flurry of oceanographic studies in the middle decades of last century. The frequency of change is rather startling. Over the last ten million years, the polarity has reversed on average four or five times per million years. The last magnetic reversal was about eight hundred thousand years ago, and we appear to be due for another one, certainly well before the Year Million. For a century and a half, Earth's magnetic field has been constantly monitored. It appears that the strength of the dipole component of the magnetic field is on the wane, steadily decreasing since 1840 at the latest, when scientists began keeping records. According to the magnetization of the minerals in clay pots made by the Romans, its strength has halved in the last two thousand years. If we extrapolate this data into the future, we find that the magnetic effect reduces to zero well before the Year Million—in fact, as early as the years 3500 to 3600.

Of course this is an estimate based on what little we know and a rough analysis of the observed trends, using our incomplete understanding of the complex phenomenon of polar reversals. Since we do not understand the mechanism causing the strength of dipolar geomagnetism to wax and wane, we cannot predict the next flip with any accuracy, especially because historical evidence reveals irregularities in the sequence of magnetic changes. During the Cretaceous period, for example, very long periods elapsed with no magnetic reversals. Further, evidence suggests that from time to time the magnetic field does not reverse itself after it collapses, but returns to its previous polarity. When the dipole effect reaches zero, Earth will not entirely lack a magnetic field, but the field will be weak and rather haphazard. And we have no idea how long this state will continue before a full dipolar field is established once more.

Will a magnetic reversal have any effect on human society and civilization? Superficially, it would seem not. There is evidence that animals use the dipolar field for navigation when migrating, suggesting potential confusion as the field collapses. However, other directional pointers can be used in its absence—for example, the position of the stars—and there is no evidence of any ill effect in the natural world. The problems for humankind occur if our future technology comes to rely even more on the magnetic field and its side effects.

The age of navigation by magnetic compass will be long gone, but will the electronic successor be any less vulnerable? The generation of Earth's ozone layer is determined by the state of the dipolar field; so when the field sags to zero the ozone layer will become unstable. A depleted ozone layer, allowing an increased influx of dangerous solar radiation, was quite rightly the big environmental scare only a decade ago. Increase in solar radiation might lead to climatic disruption, with some inevitable effect on civilization. Changes of atmospheric constituents due to industry might also combine with these effects and produce problems, unless humans can use technology to find solutions. Climatic disruptions caused by pollution and increased solar radiation could interfere with the natural navigational processes of animals and birds. We cannot forecast what effect such global disruptions will have on human society.

As time goes on, we shall gain a better understanding of the physics of the Earth and will predict such events and their political and technological fallout with greater accuracy. Whether humans possess the will to do anything about it is a different matter.

The Ultimate Random Event

In recent decades, blockbuster movies emphasized one possible doom for humankind that might strike us long before the Year Million: a dinosaur-killer rock from space.

An interplanetary body—whether a comet made of ice, a meteorite of porous and unconsolidated material, a chondritic meteorite of hard crystalline rock, or a sideritic meteorite of pure iron—gets caught up in the complex interplay of gravitational forces within the solar system, tugged by celestial accident toward Earth. Suppose it is drawn in on a curve, rather than hurtling straight at the Earth, entering the upper atmosphere at an angle of about forty-five degrees. Its speed is between eleven and seventy-two kilometers per second (25,000–160,000 mph). The friction of the atmosphere burns up the outer surface, and perhaps breaks the body apart.

By this stage, any smaller body would evaporate. Our hypothetical body, though, is big enough and compact enough to remain coherent, and a significant portion penetrates the whole atmosphere. It retains enough momentum to strike the surface with sufficient force to form a crater. The body and the Earth's surface at the point of impact are instantaneously compressed, producing extremely high pressures and temperatures. The shock wave disintegrates the incoming body and blasts out a surface cavity many times greater than the size of the fallen body itself. The cavity opens and collapses in on itself, sending huge volumes of rock debris into the sky and over the surrounding countryside. Ejecta spreads over the nearby landscape in a continuous blanket, or proceeds farther away in a rain of finer material containing the occasional large, incandescent lump.

When it vaporized the body and the impact surface, the collision's kinetic energy was converted into thermal energy, which now forms a searing plume that rolls upwards as a fireball; radiated heat incinerates anything combustible in the neighborhood. The shockwave travels through the underlying rock, causing tremors under a limited area, and through the air, compressing the atmosphere rapidly, damaging or destroying everything in its path. The rising fireball produces a region of low pressure below, causing powerful ground-level winds that suck loose debris back into the zone of impact.

Such an impact is twice as likely to take place at sea as on land, given the proportion of the Earth's surface that is covered by water. If the water is deep enough, it may absorb all impact shock and produce no seismic effect and no rocky ejecta into the atmosphere—but it will produce a massive sea wave, or tsunami. The physics of such an event, which are still being studied, suggest that when such a tsunami reaches the shore it will indeed be of Hollywood proportions. Luckily, counterevidence suggests that the energy of a truly massive tsunami will dissipate when it hits the edge of the continental shelf, and will produce relatively few calamitous effects in inland areas. More work must be done before we can predict such events accurately.

Even so, we can anticipate a widespread collapse of the continental slopes around the edges of the continents. The impact at Chicxulub in Mexico at the end of the Cretaceous period—the famous dino killer—triggered the most widespread series of underwater landslides on record. Worse, the loosening of all this sediment could release vast quantities of methane currently trapped under pressure in the form of hydrates, and so change the composition of the atmosphere. While there is no firm evidence that this particular horror actually befell the world at the end of the Cretaceous, the potential effect of such an event would be of such a scale as to give a plausible explanation of the sudden sterilization of the oceans at that time.

It remains a dire possibility for the future. Absurdly improbable? Not on a scale of a million years. Meteorite strikes happen all the time. Geological history tells us so. When the Chicxulub impact wiped out the dinosaurs, it took out 76 percent of all animal species alive at the time. So such an event would have a critical effect on our planet, and it is not a matter of *if* but *when*. Within the next million years? We cannot tell. But advanced technologies might be able to scan the skies with great accuracy and identify such "dinosaur killer" bodies long before they reach us, and divert their course. (Assuming civilization persists, of course.)

The Extinction of Homo sapiens

The above is a collection of geophysical nightmare scenarios. Nothing but change—sometimes slow, sometimes sudden. The real disasters occur when two or more of these events overlap, reinforcing each other. That seems to have been what happened at the end of the Cretaceous period. The meteorite impact at Chicxulub seems to have coincided with the eruption of the basaltic lava trapps in Deccan in India. Today we have an extra factor added to the risk. The ultimate disaster—extinction—might now befall us if just one of these events coincides with the damage already wreaked by human interference with complex natural equilibria.

So, if any of these events were to cause the extinction of

Homo sapiens, what then? Certainly, whatever the intensity of the event, it would not kill off *all* life. Something would survive into the Year Million and beyond.

Aside from wildlife in the most sensitive areas, the first victims would be the animals that owe their survival to human civilization: the domesticated varieties. Farmed animals such as sheep, cows, goats, pigs, llamas, camels, yaks and the rest would no longer be sheltered by human care. They might be so removed by now from their natural state that they are unable to survive on their own. Even today's conservation success stories would be reversed. Species once almost extinct, brought back from the brink, such as cheetahs and Californian condors, need continued human input to keep them going. They have not yet returned to a self-sustaining condition. In the absence of humans, they would drop back toward imminent extinction.

What of pets and ornamental animals, the dogs and cats? Curiously, these pampered animals may be better placed for survival, although not in their familiar forms. Poodles, dachshunds, greyhounds, or any of the other specialized breeds would not survive, but within a few generations natural selection will reinstate something resembling their wolflike ancestors. What else might survive and thrive? Mostly creatures such as rats, rabbits, moles, crows, and seagulls—animals we regard as pests, animals we cannot wipe out, however hard we try. These are vigorous and adaptable creatures able to provide the breeding stock of whatever comes after us.

As at the end of the Cretaceous period, it will be those small and unspecialized animals, eking out a living in a variety of situations, that find their way to the future. With big specialized beasts wiped out in a planetary catastrophe, they will expand and adapt, and fill the vacant niches. If the big grazing animals become extinct, then small omnivores will undoubtedly grow larger, adopt the specializations for that way of life—the complex cellulose-processing digestive systems, the long running legs for fleeing across open grassy plains, the eyesight and hearing that would detect danger from a long way away. In fact gazelles and zebras would evolve all over again—or at least animals that look and live like them—but from totally different ancestral stock.

The imminent mass extinction that might see the end of 15 to 37 percent of all species, as noted, will still leave at least 63 percent of species as survivors. Even if we reduce this figure by our own actions, in ways we cannot imagine at the moment, plenty will still be left from which to repopulate the world. Let us not exaggerate our abilities. There is no chance that we will wipe out all life on Earth. Life will find a way to survive.

Oddly enough, our past intervention will ultimately aid the future survival of many other species. Victorian natural history books made much of zoogeography—what animals live in what parts of the world and why they remain there. Kangaroos and wombats live in Australia but not in Southeast Asia; sloths and kinkajous live in South America but not in North America, and so on. There are a small number of zoogeographic realms, each of which contains its own fauna, separated from the next by barriers such as mountain chains, deserts or oceans. Once *Homo sapiens* spread out across the globe, all sorts of other species went along for the ride, and often thrived well beyond expectations.

Rabbits were taken to Australia as a food source, and they exploded in number to epidemic proportions, savaging a vulnerable, delicate landscape already damaged by crude European farming methods. The same thing happened with goats on the Pacific islands. Rats have been inadvertently released in places where they did not previously exist. And so the boundaries and distinctiveness of the zoogeographic realms have become vaguer, with a zoological sameness spreading throughout the world at the expense of locally adapted species of fauna. The hardy species—those adaptable enough to survive a mass extinction—are the ones being spread this way.

A rarely noted result of a mass extinction is the immediate impetus toward evolution. Ecological niches do not remain empty for long. Unexploited habitats provide an invitation to evolutionary development. If there is a food supply available, then something will evolve to take advantage of it—that is a fundamental rule of evolution. A whole range of new animals suddenly appears ("suddenly," on an evolutionary timescale, would mean within our one million years), as happened in the "Cambrian explosion." Delving into the early Cambrian, we find all sorts of weird fossils, the likes of which never appear again. The animals these represent lasted a few million years, until the whole assemblage was whittled down to a handful of successful lines that continue today. See, too, what happened after the end of the Age of Dinosaurs. All sorts of odd mammals and birds appeared before settling into the familiar modern groups. It is as if evolution involves trying all sorts of different possibilities, which are winnowed by natural selection to yield what actually survives in the long term.

So perhaps after a great extinction omnivorous rats would diversify into particular niches. Without the intervention of globe-trotting humans, the old zoogeographical realms might re-establish themselves, with bird-hunting rats appearing in the Palaearctic zoogeographic realm (Europe and northern Asia) and grass-chewing rats evolving in the neighboring Ethiopian realm (sub-Saharan Africa). At the same time, rabbits could evolve a similar penchant for grazing the grasslands, selection pressures producing the long running legs that we see in modern grassland animals. Competition between the grazing rabbits and the grazing rats would end with one ousting the other, and so after a significant period of experimentation the family tree of the worldwide fauna would settle into a number of stout branches with the more experimental twigs dying off.

If Homo Sapiens Survives to the Year Million

The survival of *Homo sapiens* into the far future would disrupt even the most basic evolutionary patterns and predictions. We need only look at our own history to see why that is so. However, the survival of our species is not inevitable. If humans are to survive a mass extinction, then we will not owe our survival to further evolutionary development. Brute natural selection, the foundation of evolution, has ceased, as far as our species is concerned. We have effectively halted evolution (in us, although not in the bacteria and viruses that predate us). Medical science daily saves millions of people—a good thing. This means that genes that might be faulty and have a detrimental effect on the long-term survival of the species are not weeded out, and remain in the gene pool, increasing proportionally generation after generation. Technology provides solutions for many of our most common challenges to survival. The human species is kept alive by its own technological fixes. This situation has no precedent in evolutionary history, so we have nothing with which to compare it. We therefore leave ourselves with a poor basis on which to make predictions.

Should a sudden, extinction-producing event befall humanity in the immediate future, medical science would probably be unable to take the necessary steps to keep ahead of civilization's collapse. If medical science does progress, it will probably do so in directions that seem grotesque to most of us today. When weak organs deteriorate early in a person's lifespan and cannot be replaced because healthy transplants are unavailable, the surrogates will most likely be mechanical or synthetic. We are nowadays familiar with heart-lung machines, dialysis units, not to mention wheelchairs and contact lenses; we already have a word for this fusion of human and machine: *cyborg* (from "cybernetic organism"). A time might come when such devices are the norm rather than the exception. Advancing technology would make them more portable, even aesthetically pleasing. Alternatively, or in conjunction, biological engineering might continue to improve; replacement organs could be cloned, grown, and grafted onto the ailing body. Or genomic nanomedicine might permit a reshaping of our traditional genetic blueprint at the level of cells, proteins, and atoms. By this point we would begin to abandon the conventional human shape, coming to accept the shape of the bundle of accessory artificial organs, unique to each individual, as an integral component of the individual's identity.

A Backward Glance

One million years in the future, the world will show a physical geography not much different from that of today. The planet's large-scale and local climates will have changed dramatically, but not the landscape: a shift to landward of a coastline by a few tens of kilometers is not going to make that much difference. (Unless, as other chapters suggest, Earth has been dismantled entirely!)

Exposed geological strata from our own time—visible only in regions of active mountain formation such as the west coast of the Americas, the Mediterranean area, and Southeast Asia, where they will rise from great depths—will form a distinctive horizon or particular layer in the sedimentary succession. Certain areas of this horizon, corresponding to cities and landfill sites, will show concentrations of glass and plastic, traces of which will still be visible after this period of time. A bed rich in lime and clay minerals representing building materials will be so compacted that no structure will remain from our once-great cities. Probably there will be a great deal of mineral discoloration—reds from oxides of iron and greens from the salts of copper—reflecting our use of these substances.

Elsewhere ocean sediments will show a deposit of heavy metals and maybe (chilling thought) a concentration of radioactive isotopes in one particular horizon.

It will remain a living world, whether or not humans are

part of it. Biodiversity will be particularly high, as evolutionary experiments will continue to take place, newly evolved lines competing endlessly with one another for long-term occupancy of the various niches.

Earth will continue to be a little oasis of life in the vast firmament.

Chapter 3

A Million Years of Evolution

Steven B. Harris

A million years is not much on the scale of astronomy, nor even of most geology. A few rivers and seas change. If continents move a hundred kilometers (about sixty-two miles), the difference is hardly noticeable. Discounting human macro-intervention, the Earth will look about the same in the Year Million, give or take a polar icecap or two.

In biology, though, a million years can work wonders—or, sometimes, do nothing. Much can be accomplished over the tens of thousands of generations this span represents for mammals, if the pressure to adapt is extreme. Two million years ago, we humans were big meerkat-style hominids, either *Homo habilis* or something similar, standing four feet high on our hind legs and shambling about on the East African savanna, sometimes falling prey to leopards. At this stage we were not an impressive animal, but we were learning slowly to handle rocks as tools. We had little other choice, exposed on a relatively treeless plain, already too large to find safety in hole digging. A million years of such pressure, or perhaps even less, turned us into hulking, six-foot-tall *Homo erectus,* with a brain now expanded to 75 percent its present size, a nearly modern human body below the neck,[1] and the ability and proclivity to chip big rocks into wicked two-sided razors. By this time we were at least no longer cat food—even if it wasn't yet apparent that we were destined to be rulers of the planet.

Then even faster evolutionary mechanisms became available. Along with a big brain came several new selection mechanisms, each of them with a drastically reduced timescale.

Consider artificial selection, an improved type of imposed evolution running at high speed. Darwin knew it well, and knew vaguely that it works especially quickly. Artificial evolution can move much faster than Nature because the interventive pressure that runs it can be maximized. Nature rarely eliminates all animals that lack a given trait. Artificial selective breeding can and does do so, nurturing survivors, protecting them against predators, and providing ample nutrition. Most of our modern agricultural crops are scarcely older than ten thousand years, and most of our domestic livestock breeds, too. That is the accelerated pace made possible by ordinary "barnyard" artificial selection. The relatively recent development of toy poodles, bulldogs, modern cultivars of maize and wheat, and other improbable organisms that would not survive in the wild shows that evolution can be engineered and turbocharged.

But did we artificially select *ourselves*? It doesn't take laboratory equipment to carry out applied genetics, and humans have always been at least as interested in our own breeding as in agricultural breeding. This is *sexual selection*. Other animals use it, to conspicuous effect. Even in the wild, peacocks have ridiculously long and survival-handicapping tails: good for impressing the opposite sex, bad for escaping a predator. Peahens evidently use a glance at the peacock's tail as a simple one-shot look into the general health and genetic fitness of a male, who is bragging that he can support growth of a superior-looking tail *despite* the pressures of parasitism and predation. Sexual selection is particularly interesting if done by high intelligence, which then has an opportunity to sell itself. In this case, intelligence itself becomes a sexual attractor.

Interestingly, the mushroom-like growth of the human brain over the last 1.5 million years—it has increased a third in mass while the rest of human anatomy changed relatively little—resembles nothing so much as that of the peacock's tail: its utility is not immediately evident. Obviously, though, much evolutionary pressure has been applied differentially above the neck. The human brain now requires 20 percent of our resting calories, and is the largest for body size of any mammal. How did we get it? In humans, language and fine athletic and other coordination (which takes intelligence) are some sexual selection factors. Such abilities provide a look at the function of the brain, an organ that is not so easy for other species to evaluate in their own mate selection.

In other animals, however, the brain is not so critical as in humans, who have few other physical gifts. The way humans survived in their savanna niche depended on keen wits and cooperation. Consider the importance (and difficulty) of assessing "team-player intelligence" in a species that has few natural defenses, and relies on the ability to coordinate hunting groups for survival. Perhaps our ancestors selected themselves into being what we would now recognize as "big players" on athletic teams, who attracted the sexual interest of the most desirable admirers—"desirable" by whatever local standards, including health, youthful beauty, charm and wit. (That happens in all hunter-gatherer societies in the present day after a successful hunt, so we might suspect it's been going on for far longer than recorded history.[2]) At the individual level of reproductive choice, men and women have always been at least as choosy about whom they breed with as we are now. We *Homo sapiens* are not like our relatively promiscuous bonobo chimp cousins. Women have always made men work hard for their favors, because a human infant is a relatively immense burden, in terms of calories and time. So human males rapidly grew the size of the parts they needed, and that included the organ behind the eyes. In turn, those changes were handed down to both sons and daughters, and so on, in a self-reinforcing cycle.

Cultural Evolution

The subtle brain differences that make Indian elephants potentially rideable, unlike African elephants, aren't visible even to a microscopist working with fresh brain tissue. But they are governed by the DNA of both species (even if we don't know which genes are involved). Even subtler changes, not genetic at all, can take place over time in an individual's brain—changes in *software*, so to speak. The behavioral difference between a wild mustang horse and a saddle-broken horse is simply a different set of programming, for one can be changed to the other. For humans, this software, which we call our "culture," has become ever more important to our behavior—yet culture, too, is subject to evolution. Humans raised without socialization turn out to be barely more than animals, despite their large brains. Those brains, therefore, represent only raw processing power (with perhaps some doglike domestication proclivities), and they need a lot of programming.

Homo sapiens sapiens is now largely a software species, perhaps the first, governed mainly by *epigenetic* factors (outside the genome), some of which are *extrasomatic* (outside the body). Much of what makes us special as a species is stored not in genes or brains, but in libraries, laws, traditions, and songs.

Artificial selection (of which sexual selection is just one instinctive part) requires some culture to decide on breeding goals and standards. How can a society decide what characteristics it desires in its mates, children, and livestock without cultural *values* for such things? Without language, there are limited possibilities for cultural variation and depth. Some culture (extrasomatic knowledge passed between generations) has been observed in higher animals, but it seems limited to a sort of direct learning via direct imitation. Not much knowledge can be stored in this fashion. Language and a good memory allows cultural programming to become predominant over genetic programming. A good brain can store many thousands of lines of oral tradition, particularly if formatted mnemonically as a saga or liturgy. And even without language, a good brain can also store a large amount of motor learning about how to knap flint, use fire, and weave textiles.

We can't tell when humans began to exchange such complex pieces of cultural programming, but somewhere in the Neolithic era (with active agriculture and large cities, starting around fifteen thousand years ago), people started doing complex tasks that must have taken a lot of coordination—far more than the average wolf pack or lion pride manages.

Our own high culture came very late. Something more was needed for the next phase in evolution to happen without any obvious brain growth. Near the end of the last ice age, perhaps ten thousand years ago, bands of people began to settle in the fertile crescent in Asia—to wait for food crops to grow (which they had learned to replant annually) and to hunt the beasts that came to eat the crops. Soon, they began working to tame and remake these beasts as they had the plants, and eventually had domestic animals and were fully farmers. Rooted from year to year for annual cycles of harvests and slaughters, they built the first cities, and in the process they invented specialization. In some cases, markers for stored goods evolved into written symbols.

Writing, when it arrived, changed everything. Now brain size and intrinsic memory were not so important. Complex contracts and laws could be impressed even into tablets of clay. Societies could now store their cultural programming outside oral tradition. That allowed (finally) for the technological leap into the present day. This leap depended on a completely new form of evolution in culture, one dependent on information storage in which most of the content is stored outside the structure of the brain. Operations on this kind of data could be done much more collectively, across both space and time.

People who cannot write also cannot achieve a culture that produces anything so complicated as firearms, much less the rest of an industrial society. An industrial revolution requires printing. Before the movable-type printing press arrived in the fifteenth century, record keeping for complex collaborations was just too much work. Printing mechanisms allowed spare time for complex coordinated activities, for experiments, and for the easy reproduction and wide dissemination of the results: science and engineering complicated enough to need blueprints and careful specs.

Finally, when electronic computers arrived and shrank enough to fit into homes and then into the hand, suddenly, almost magically, individuals gained digital access to *all of human culture.* That assimilation process seems likely to end within a generation.

Culture as Artificial Intelligence

Our next limit in a fully digitized culture (now fast approaching) is total available processing power, rather than access to the information to be processed. The next limit is therefore the connectivity and bandwidth of communication between brains. Most people can talk (and write) faster than they can produce first-rate thinking and conclusions. So while brain-to-brain connection remains limited by the information transfer rates of speaking and writing, that's not nearly as big a problem as it might seem. Until now, flows of high-quality information in and out of brains have been crimped more on the input side (such as looking for information at the library) than in the output side (writing content worth publishing and looking up in that library). Fortunately, the worst crimp—the search problem—is disappearing fastest.

Human culture is mostly software, but a society is more; it's partly interactive software that does information processing (by both brain and computer), and is often unaware of what it's doing. Adam Smith realized that economies run that way, using what we'd now call "parallel decentralized processing"—a process where nobody in particular is in charge, and there is no central control. In one sense, that's some of what our high culture is: a collection of increasingly intelligent "entities" constructed out of multiple brains and data storage centers.

Such collective entities can do things no single human can, and this was true before computers. A modern city or a nuclear aircraft carrier is beyond the mental capacity of any one person or group of unaided persons, but it gets built anyway, through the collaboration of thousands of engineers and whatever devices they use to communicate and draw plans. Such megaprojects already depend on a form of "artificial intelligence" (intelligence greater than that which humans can mount without artificial aids). Today, the augmentation of intelligence needed for such tasks is mediated by data processing machines. Does it matter whether or not *some* of the total processing to build an aircraft carrier is still done organically? That fraction done by brains will finally become irrelevant, when the distinction between processing information organically and "electronically" disappears.

Moore's Law and Accelerating Change

Electronic data-processing power per unit of money is increasing exponentially and shows no sign yet of hitting any wall. At some point in this century, information will be processed by manipulating spins of individual electrons on appropriate electronic "devices." Some number of these (I won't guess how many, but I think it will be an attainable number like a million or a billion) will equal the processing power of a human neuron. This must be a recognizable number because a neuron itself contains around five hundred trillion atoms or so (this figure is calculated easily: most are atoms in water molecules), and most of these atoms aren't harboring a spinning electron involved in some neuronal calculation.[3] It has been suggested that neurons are quantum-entangled computers, but very few people who know of the jiggling that happens in our warm interior world of cellular proteins believe it. No matter whether neurons are complex computers or simple ones—analog, digital, or something in between—at some point it will be possible to replace any given neuron with its electronic equivalent. Then more of them. Then finally (why not?) all of them.

Well before then, however, it should be possible to connect neurons to neurons outside their native brains. At some point, mechanical telepathy will become possible, with all that this implies. Imagine being able to remember the entire contents of our digitized civilization with the same ease that you remember basic spelling. Now, imagine being able to do that trick with selected memories from another human being. Scary thought, isn't it? Do you trust anybody enough to give them access to your candid diary, let alone your entire memory files? Possibly not.

But suppose you did, for a while: when you "pulled the plug," how much of the other person would remain "in you," as a part of you, and vice versa? That would depend on how good your file copy system is. If it were large, you could retain copies of the other person's memories to access at your convenience. At some point, you could even give these borrowed memories permission to influence your unconscious actions. You would then acquire not only something of another person's declarative memory, but also the other's intuition, judgment, anxieties, and passions. Already, people in a long and close relationship often know what the other is going to say. This sort of internal emulation could be amplified to any extent.

Now imagine drawing in other minds, perhaps of anybody willing to participate. The result is the Internet on crank: direct brain-to-brain connection, amplified by computerized search and processing help.

The Chain Reaction of Intelligence Explosion

Explosions are typified by runaway positive feedback. A familiar example is a chemical explosion, in which the reaction releases heat, which in turn speeds up the rate of the reaction, liberating more heat. This leads to an exponential growth that continues until it runs out of fuel, or is otherwise stopped.

All the kinds of intelligence increase we have discussed have a built-in positive feedback loop, starting with our first homely example of sexual selection for intelligence. Presumably, more intelligent animals become better at *any* kind of sexual or artificial selection—including those that produce more intelligence. But what's special about selection focused on intelligence itself is the positive feedback loop that makes the process accelerate. This exponential curve probably explains how human high culture arrived on the scene so fast—after those hundreds of millennia of chipping rocks and wearing skins. Only recently did we reach the point where a certain spark has received the fuel to create a "blow-up" fire.

We know it's possible for humans to vary greatly in intelligence without much difference in brain size. Culture matters. For example, there's a well-known "secular I.Q. trend" (the Flynn Effect) that extends across many cultures and socioeconomic groups, indicating that as a species we gained more than twenty-five I.Q. points in the previous century. Nobody really knows why, but it doesn't seem to be better nutrition. Instead, it seems we've simply learned to think better, with better educational tools. Thus we see directly that intelligence can improve by strictly cultural means, even if hardware (wetware) doesn't.

The rapid exponential growth of human intelligence becomes inevitable once it is freed from the constraint of having to wait, generation by quarter-century generation, for better brain *hardware.* Cultural evolution works fast. All the positive feedback loops in *effective* intelligence are present for cultural evolution, but the cycle time shortens and reshortens, depending on ease of information storage. Our brains probably haven't improved much in innate capability since the last ice age. But with information technology driving "effective intelligence" progress (measured simply by gross capability), we have moved from writing to the printing press in 8,000 years, then from press to mimeograph in about 430 years, from there to the photocopier in less than 75 years, to personal computers and universal word-processing in another 30 years. That takes us to the late 1970s. The modem reached ordinary consumers in the early 1980s, the world wide web software that now runs the Internet arrived in 1990, and the first workable web browsers for the masses, about 1994. Ten years later, these became portable, through wi-fi and cell phones.

As I write (2007), a billion of the world's 6.6 billion humans have regular access to theInternet. Another decade or two should see this particular level of interconnectivity—only about thirteen years old now—nearly complete, save for some backwaters. Even these might be surprising: cheap cell phones have penetrated to parts of many countries where people have access to a phone but still not to clean water or enough food. This trend will continue with web-enabled cell phones that are now expensive, but will not be for long.

When the average person has access to the Internet, all culture will have moved accessibly outside our brains, although the ability to process it will mostly still reside within the organic parts of the machine. There we hit a temporary snag, but even after full culture digitization and full downlink connectivity, we'll still have 997,000 years to go before the Year Million. Patience.

In retrospect, it is odd that we didn't fully realize until about 1965 that the positive feedback loop leading to explosive intelligence increase was capable of operating overtly. We were looking at intelligence in the wrong way, not realizing that you didn't need full artificial intelligence (AI) to get intelligence augmentation, and consequent exponential evolution in intelligence. In that year, I. J. Good finally addressed the basic question of runaway feedback, starting from a strictly AI viewpoint. Good asked what would happen once computers became powerful enough to completely design "smarter" computers (he didn't ask what would happen once computers got smart enough just to *help* design computers). Science fiction quickly suggested thinking machines far smarter than humans, mostly malign: the computer in the movie *Forbin Project* (1970) that we'd like to unplug but can't; and *Terminator*'s cold and lethal Skynet (1984).

None of these prospects, however, are fundamentally different from the feedback loop in intelligence raising we've been engaged in for millions of years. What is taking place now is a new leap-after-leap forward in the *pace* of the loop cycle. So we're noticing it more and more. The next big metajump will take place when human brains talk directly to computers, and to each other.

Direct Connection of Brains

We are not smart enough to do this yet. We're still at the level of sticking electrodes into brain slices and reading the waveforms. But we can guess what will happen when we connect brains more directly. Composite minds more intelligent than any single person are not only possible, but something we've already experienced. They impart to our high-tech culture some of its odd alien nature, from the economy to the Internet, as though it were all controlled by some godlike conspiracy that hasn't let us in on the big secrets. But of course, that isn't true. The composite and distributed "mind" that runs economy and culture is indeed godlike in its abilities, but there's no conspiracy. It seems secretive because it doesn't talk, but that's only because it *can't* talk. It's not self-aware, and nobody with full understanding is "at home" inside it, in the same way nobody who understands chess is "at home" inside a chess-playing computer. Nevertheless, the end result appears creative and understanding, even to a chess master. Such a system does its job in a massively parallel way, without a central government, just as Mother Nature does. It works very cleverly, but it doesn't explain itself or its thinking (in fact, when parts of it try to do so, we see how ignorant they appear, and wonder if they are part of the conspiracy). We ourselves are pieces of one or more giant "cultural minds" and make up parts of them, but none of us is really aware of the total process, any more than a shopper is aware of participating in a vast economic calculation.

While he was developing his theory of organic evolution, Darwin deliberately read economist Adam Smith for help. He knew he needed a theory in which *design* happened "all by itself," without any central planning department. And indeed, all types of evolution, as we now know, are basically equivalent. Living organisms were "designed" in many cases like Wikipedia articles. The process is gruesome, like what Otto Von Bismarck said about laws and sausages (i.e., those who have any regard for them should not watch closely as they are made). In all evolutionary processes there are plenty of bizarre and vestigial bits left over, and some malfunctions and vandalisms. But after enough time the overall result, as with the economy and the common law, can be quite magical, and certainly better and more creative than anything top-down or centrally planned. That is the paradox.

What is the experience of being part of a superintelligence? We already know some of the answer: like being part of a team doing an important task. To everybody on the team, it feels as if godlike people are doing most of the work someplace else. But if you look around, you can't find them. Our experience residing in our own brains mimics this experience to a degree, because most of what we do, and think, goes on below the surface of our conscious awareness. We get some hint of it in dreams, when the parts of our brains engaged in modeling the actions of other people suddenly split off and take control, acting as the bizarre stage builders, playwrights, and actors of the little dramas we find our "dream selves" moving through.

The Borg and Assimilation of the Organics

Some say the next tipping point to the evolution of mind will be I. J. Good's singularity-like scenario of "ultraintelligent machines"—the point where we can directly redesign the brain (or any brain) to be more intelligent than it was. This enhanced brain is then smart enough to design still better brains, until the whole self-bootstrapping process leaves us behind. This might be too optimistic: the programming and design of such artificial brains is beyond us, and might require evolutionary help from self-evolving software. If so, AI specimens of human power may not end up understanding themselves any better than we understand ourselves. Instead, all we will have achieved is substrate and programming that can be improved more directly and rapidly.

The grossest capabilities of a single organic brain can be changed now only by artificial selection, and thus human capacities won't grow much in the near term—not until we have full nanotechnology, and the ability to build and replicate brains from the cells up. Long before then, however, connectivity between brains and interfaced computers will vastly increase our effective intelligence. At some point it won't matter much whether computers are smarter than humans, because what others have called the "Borganism" (the Internet of brains and computers that directly link them) will reach the capacity to do the next step, which is to control and grow computational matter. Then, organic brains as we know them (and grow them organically, by unskilled labor) will no longer be needed. That shouldn't be too scary a thought, because the "people" who inhabit such collective thinking organizations will still be part of the whole conscious entities. But these people *will be programs*, "running" on substrates other than gray matter.

With old-style organic brains removed from the Rube Goldberg machine of animals and devices that had been the interconnected network of mankind, at last we will have hit the final evolutionary phase—a pure expression of what we've been calling pure "cultural evolution." Ultimately, the evolution of software, ideas, and knowledge is the only evolution that counts on short time scales. Any software or hardware design innovation can be translated eventually (if you have the technology) into thinking "hardware," without waiting for a present-style "organic" brain to grow in the womb, be born, and be programmed. Those scare quotes around "hardware" (or "wetware") are needed to remind us that *computronium* (the material used to do the information processing) need not be silicon chips, and almost certainly won't be. How mechanically "hard" computronium need be is also open to question. Perhaps it will be diamond compiled from carbon atoms, or a diamondoid sponge. Certainly the mobile units that connect with it when physical movement is required won't be metal robots that clatter around. Why build such awkward things, when we don't need to?

While we don't yet know what the mobile units will be made of, we can imagine that the distinction between living and dead matter will lose its meaning. Any technology capable of copying living organisms is capable of copying them with better and tougher components (diamond-fabric woven bones, for example), and any of this material can justly be called "alive."

Probably the new "stuff" that results won't (or will rarely) consist of anything so delicate as twenty-micron-sized bubbles of water, full of protein and DNA, scaffolded and bound to each other with stringy polymers, as though somebody were packing water balloons individually with fishing line and duct tape for long-distance shipping. That's an accident: life started as little balloons that floated around in water and didn't need much support. When finally these squishy cells needed support to resist the gnashing of predatory teeth and the vagaries of life on land, they had to develop it by bricolage, and couldn't redo any of the basics en route. Won't it be nice to get rid of all of that fragile stuff? The present human body is like a bicycle made of porcelain, which mostly can't be repaired; all evolution has done is add more packing.

Bodies of the future will not only be much tougher, but will also feature backups of brain information in case of total destruction. In fact, the primary thinking centers probably won't even be located in the mobile unit/body, so backups of memory and thinking will be automatic. There is no physical reason any of this change need *feel* any different, to the user, from the body being "worn" now. In fact, it might well feel like an improvement, with enhanced senses and abilities. What feels "natural," when examined closely, is strange enough already. Remember, your present brain sits shock-insulated in the dark, and gets only a view or feel of the world by means of chains of digital signals coming in from sense organs. All this can be manipulated—in theory—at will.

Computronium: Software in Action

The actual structure of computronium is unimportant. Mobile units might be soft and warm and, if desired, have a metabolism. One might even try for the complete "look and feel" of the human body, but without delicate cells full of water, making the unit extremely durable. There's no reason such a mobile unit shouldn't have a processor capable of individual thought (perhaps of a better grade than we currently enjoy), even when disconnected from *everybody* else (the net). The shape could be tailored to any occasion. For exploring the ocean, we might want something like a squid; for the air, something like a bird. On land, for a change, one might choose to be an efficiently running four-footed body modified with extra arms and hands, like a mythical centaur. The same applies to exploring other worlds: if the mobile fits, wear it. Whatever can be designed, it will deserve the title and privileges of "human being." When we look out of the "eyes" of such bodies, we'll certainly still think of ourselves as "human"—if we remember our heritage.

For those of us who want to play at being totally human in form, why not? The Amish of the future can choose to use the outhouse if they like, and can choose to *need* to use the outhouse. If that makes an organism feel organic and "creaturely," well, it's available. Remember, however, that if you have a fully human brain in your chosen "body" there will be limitations to thinking, perhaps to connectivity, and a certain physical frailty—so don't complain if there are games you can't play in that mode.

The interconnected mechanical/organic composite Borg (short for *cyborg*, or cybernetic organism) of *Star Trek* is horrifying for several reasons, but one is that they seem like badly patched-together and ugly blends of organic beings and hard computerized robot gear.[4] In any case, we can dismiss the television horror view—that's what things might look like at worst, when brains are still part of the equation, with no way to replace them. Even then, nonorganic or biological prosthetics for brains are likely to be considerably less repulsive than a deliberate monster show. By the time nanotechnology is capable of building complete bodies, using computronium for thinking and other morphable substances for moving, nobody will want to look like the Borg—except for a Halloween party. Instead, a body will look and feel however you want it to, sensitive but tougher than flesh.

And of course, since such stuff should be infinitely repairable and connectable, it will be effectively immortal. No material object is indestructible, but aging as a problem simply goes away once brains are fully inter-connectable, because whenever a part (even a thinking node) is damaged or wears out, consciousness moves to another location until the damaged body or part is replaced. Repair of bodies (and "brains" or processing gear) should be like replacing any backed-up computer node in a networked system.

Individuality or the choice of solitude in such a world will continue to be an issue that people will need to solve in their own ways, probably determined by available bandwidth of communication, with all that this implies. Choose a wide bandwidth, and find your consciousness completely immersed and diffused in whatever number of others with whom you've decided to merge. Chose a bandwidth of zero (or have it forced upon you by distance or equipment failure), and you become once more, in that sense, an individual. By the Year Million, you might not like such isolation. Or you might be the sort to do it deliberately, as a survival challenge. No doubt the future will contain many such games.

The Singularity

At some point, organic brain material will be replaced finally with computronium. As noted, the historical juncture at which designed mental substrate becomes as complex as the human brain has been termed the *singularity* by mathematician and writer Dr. Vernor Vinge, who coined the phrase in 1982, following the 1965 lead of Dr. Good. Intelligence increases in machines wouldn't cease, for it remains exponential and finite, but it would change its cycle time for improvement—to what cycle time, no one knows.

The point of the term *singularity* is that all intelligences past this point are both superhuman in scope and capability (as in fact is any working group of humans *now*) and, in addition, such intelligences presumably have the capacity to improve their own effective intelligence more directly than we have now. Their *own* operating code, unlike ours, might be an open book to them.

Extrapolating the complexity of computer systems, compared with the estimated complexity of the human brain, the timing of the singularity has been estimated by many futurists to arrive before the middle of the twenty-first century—far short of the Year Million. But are there likely to be any fundamental growth limits that might effectively stop super-intelligences from simply taking over after say, 2050, and then doing anything they want to do, omnipotently?

Limitations of Energy, Time, and Space

There must be limits to intelligence—because time, space, and matter limits apply to the substrates used to create intelligent matter.

Let us assume modestly that in the future there will be no "magic physics." The usual conservation laws of physics will continue to hold, thermodynamic limits of entropy are real, and causality continues to be inviolable, so that reverse time machines are impossible and the speed of light remains a limit. If we abandon those rules, prediction becomes uninteresting, for then anything becomes possible. I also assume for our purposes that dark energy and dark matter will turn out to have strange properties and strange uses, but none that violate any of the aforementioned laws and properties of the universe.

What, then, are our limits as we move toward the Year Million? No matter what body forms we use, we will need energy and matter. We might be able to partially take apart the Sun, using its own power, but that's a losing game, since if too much mass is removed, the power fails. Matter obtained this way is costly in energy, since it is deep in a gravity well. Removing even the outer 0.1 percent of the Sun (assuming perfect efficiency) requires thirty thousand years of solar output energy. This mass is about the mass and composition of Jupiter, which can be used for the same purposes at only 1 percent of the energy cost. So the outer solar system will be mined for material before the outer Sun is.

Jupiter represents more than 70 percent of the solar system's non-Sun mass, so the entire system can be considered, for our crude purposes, to consist *only* of Jupiter and some other debris (like our present habitation) that isn't worth bothering with in our calculations. For example, excluding the Sun, most of the solar system's energy of motion is in Jupiter, as is most of the gravitational energy of self-binding. So the other debris will presumably be used for practice when we need material, or (as in the case of Earth) set aside as memorial, if we choose.

The more intriguing part of Jupiter appears to be its supercompressed core of carbon and heavier elements, probably about six times the mass of the Earth. We probably need heavier elements than hydrogen and helium to make computronium. Hydrogen itself as a "thinking material" looks a little iffy, because it doesn't seem complicated enough to make anything useful with it. Ultracompressed hydrogen probably turns metallic, but that doesn't help us much, either: electronics as we understand it requires electrons to be confined and isolated, and in metallic conductors they are too free.

In any case, we'll build clouds of solar-orbiting collectors out of the heavier elements, in order to catch all of the Sun's light. Powered by this energy, our information processors will be placed at nodes somewhere within the cloud. Such a cloud is called a Dyson Swarm, after physicist Freeman Dyson, who first envisioned this engineering feat. For computation, we need nodes with dense processor connections, because otherwise we have speed of light connectivity problems. However, we can't put the entire computronium collection in one place, because its gravity will cause it to collapse; computronium cannot be a simple material, so it will have difficult pressure limits. Also, we have a lot of heat to get rid of, because it won't be perfectly efficient, and this requires a complex structure for forced convective cooling. Both of these things will dictate a mix of nodes and connections. [Robert Bradbury's chapter addresses the tradeoffs of problems in making Matrioshka brains, or Mbrains, which are planet-mass onion shells of distributed processing nodes associated with clouds of solar collectors around light sources such as stars and fusion reactors—Ed.] But imagine an orbiting cloud of collectors and processor planetoids, all bound together by coherent beams of connecting radiation. Such a cloud would orbit a star or any smaller fusion source that puts out light.

To make the largest of these, dismantling Earth would be just the beginning (or Earth could be preserved while we begin with asteroids and the smaller planets). Here's a motive for pulling Jupiter apart (provided we don't find exotic life there, floating in the clouds). How much energy of self-binding must we overcome to do it? It's convenient to calculate it in terms of solar output. It would take twelve years of solar output to move Jupiter out of the solar system, but about three hundred years of output to disassemble Jupiter completely into usable gas (other authors, like Freeman Dyson, have figured this at a few times higher, but we need only order of magnitude estimates here). The figure for disassembly of Earth is about six days, indicating why we can ignore it, and even the other gas giants that are only a tiny fraction of Jupiter's usable mass. "Jupiter" stands for everything that is outside the Sun's gravity well.

The Sun provides energy at larger rates than most other processes we can think of, so its energy will likely be tapped first. That rate is 4×10^{26} watts, and given the thermal gradient from the 5,500° K (degrees Kelvin) sunlight to the 2.7° K of the microwaves in interstellar space, nearly all of this energy is available to do useful work, in theory, depending on how clever we are in building solar collectors. And of course we will be very clever. We have only about eight to ten Earths' worth of non–hydrogen/helium building materials in our entire non-Sun solar system (most of it inside Jupiter, as noted), and this may not be enough to collect most of the energy of our Sun. For example, at the radius of Earth's orbit, a shell of this matter would be only a few tens of meters thick. If it were to be converted to orbiting clouds of solar-power satellites (even disregarding what matter we'll have left over to use for thinking), we end up with effectively a very thin layer of matter to stop and process the Sun's light. We may not have enough for really good efficiency, but it will be a good start. To get a limit of 90 percent efficiency we need only radiate waste heat to space at a temperature of one tenth that of the Sun (at 550° K), and the collector orbital radius to do that is well inside Earth's orbit—something like the orbit of Venus. This requires only half the collector material.

What if we need more energy than this? Or if we need the energy from solar collectors, but cannot afford material to build them, because we need it for computronium? (It would be nice if certain chemical elements work for collectors and not for brains, though we can't guarantee this.) The Sun produces energy only at a certain rate, and if we want to get energy faster—perhaps for building space probes to shoot at the billion-plus stars within ten thousand light-years, or merely to power faster thinking in our own solar system—we'll need to extract it faster. Remember, by this time we will think really fast, and we will be ready to use energy much faster than our puny star releases it. We can always go to other solar systems to build more collectors for our star, but that's only going to raise the output a fraction, because we already expect reasonable efficiency in our "Dyson-swarm" collectors. If we must go space-faring to get energy, there are better alternatives.

If we want to disassemble the planets for their elemental content, we're going to look closely first at their fusion fuel content.

Mining Deuterium

We want Jupiter and the other gas giants not only for the heavy raw materials at their core, but also for their deuterium. Deuterium is the heavy isotope of hydrogen absent from the interior of the Sun, because it's destroyed there. No natural processes create deuterium in significant amounts today, but it was produced in the Big Bang and so remains in the Sun's outer layers and in the gas clouds that form planets in our solar system and other star systems. However, it's a diminishing, nonrenewable resource. All the time a star burns, or when it goes supernova, scattering the ash that makes new star systems, the process destroys some of its deuterium. As time goes on, less and less deuterium is left in the universe.

Deuterium is very much easier to fuse to get energy than is ordinary hydrogen—so easy that the first true U.S. hydrogen bomb was little more than a big thermos bottle of liquid deuterium, squeezed by an ordinary fission bomb. At ten megatons yield, this worked fine.

On the materials side, about 98 percent of Jupiter is hydrogen and helium, which we can perhaps find some use for, as computational material. But even if we cannot find any practical way to use hydrogen and helium as substrate computational materials, we can possibly use them as reaction mass for our fusion rockets (Orion, etc). And there is always the remaining 2 percent of Jupiter, which is still six times the mass of the Earth and represents much of the usable heavier elements available in the solar system. (We don't yet know the exact mass of the Kuiper belt and Oort clouds, but there is reason to believe it is small enough to ignore.)

As for using Jupiter's ordinary light-hydrogen (not deuterium) for nuclear fusion power, we're far past the edge of what is technically possible there. So far as we know, ordinary matter and electromagnetic fields cannot confine this light isotope well enough to get much energy out of it efficiently, so only gravity works for the process. We know the light isotope of hydrogen *can* be burned in fusion, because stars do it—but on the other hand, nature might *require* a star for it. A clue that this may be true is that, even at the heart of the Sun where temperatures are fifteen million degrees Kelvin and densities are far higher than can be sustained indefinitely by any other means, the energy output is a modest 15 watts for a volume the size of the adult human body—a reptilian metabolic rate about a fifth of the energy production of the human body itself.[5] The Sun manages its prodigious energy output with the poor fuel it uses only by having a vast volume and by working continuously—two things that cannot be done simultaneously by any imaginable device.

For this reason, humans at present cannot fuse light hydrogen, in either bombs or machines, and we have no idea how to do so artificially. It's just too difficult. It's one of those things that might be possible in the future (like making small black holes)—possible in theory, but it might never work out. But it's a worthy goal. If it *can* be done, it's worth doing, because Jupiter has about one one-thousandth the mass of the Sun, and thus holds enough hydrogen to equal at least eighty million years of solar output, in fusion (compared to the eighty billion we could in theory get from the Sun's hydrogen; though even the Sun is expected to get only 15 percent of this before it goes haywire).

We know Jupiter has deuterium (twelve parts in ten thousand of its hydrogen by weight). Dismantling the giant planet for energy is worth doing even if the deuterium in Jupiter is the only usable thing we find (we could fuse deuterium even with 1950s' technology). Still, Jupiter's total deuterium supply is equal to about 30 percent of the mass of the Earth. We can get only about twelve thousand years of solar equivalent output from that source.[6] Still, it's more than enough to pay for mining it out of Jupiter's gravity, because by using solar power we can take Jupiter apart in a few hundred years, and burning its deuterium saves us having to wait twelve thousand years for the Sun to generate as much energy as we get from the deuterium. The ratio of energy obtained from Jupiter's deuterium to the energy needed to pull Jupiter apart to get it is at least ten to one.

Deuterium energy, as noted, is easy to harness. For bombs, the fission trigger that does the initial heating and squeezing can be made of minimal size (a few kilograms of fissionable material), after which successively large deuterium bomb stages can be used as igniters for larger ones, ad infinitum (three-stage bombs have been made already, and there appears to be no limit to how many stages a bomb can have). Thus, we won't run out of uranium before using all our deuterium.

The energy in deuterium can be used directly by surrounding a bomb with solar-type energy gatherers, or else a series of small bombs can be used as a pusher in an Orion-style spacecraft. Presumably our superintelligent progeny will also at last figure out how to make "slow-burning" deuterium-based thermonuclear power plants. This is something we haven't quite done (at least efficiently), but being well on the way to it we have every reason to think it's possible, as might not be the case with light hydrogen (even in the Year Million).

What will we want all this energy for? Pure computation in theory costs zero energy, but storing the results and memories does, and we might need the energy from deuterium fusion for that. This may allow us to use the eight or ten Earths' worth of non-gas material in the solar system to build things other than solar collectors.

If it's possible to fuse light hydrogen, deuterium collection won't be worth doing—we'll have a surfeit of energy just from Jupiter's stock. But if we can't do this, it might still be worthwhile to visit local star systems to harvest deuterium from the many gas giants that must surely be circling nearby stars. For example, some sixty-six stars reside within ten light-years of Earth. We believe gas giants are common in star systems, since astronomers see a lot of them. Even if the sixty-six star systems at near distance hold only 10 percent of the gas-giant planetary mass that our system does, each of them would still have (on average) one thousand years' worth of our total solar output in deuterium stocks, or sixty-six thousand solar-output years, total.

Boosting this much gas to 1 percent *c* velocity, for transport between star systems so we can get it within one thousand years, is actually not very expensive energy-wise. At good efficiency, perhaps using linear accelerators, all the gas-giant deuterium orbiting a Sun-like star (twelve thousand solar-years output worth) would cost less than a thousand years of stellar output at the "mining star" to send our way. It should be worth doing, even with the inefficiencies involved (remember, if we use a star to power a laser for a solar sail launcher, then we're effectively using a working light-hydrogen reactor to ship heavy hydrogen to us). We can slow it down at this end via solar sail, by using solar-powered coherent beams of radiation, at the same cost. Think of the future as involving streams of deuterium coming into our system from other nearby star systems, something like water converging from all over California into Los Angeles. Let us hope there are no aliens to object to this redistribution of galactic resources.

These issues are framed in terms of bringing resources home, because there's no point in going to other star systems if we're going to stay there and use resources that way (see colonization issues below). The whole point of collecting energy from other stars and sending it back to a central place for use is that it can power a massive computational system that is compact and not limited by connection delays at the speed of light. So long as light speed remains a barrier, people in other star systems will remain forever individuated by their distance: they won't be "us." If we want to use other stars' energy for thinking, the energy will have to be sent to our solar system.

As our probes travel outward, converting the gas giants of other star systems into deuterium mines that are sending a stream of energy home, the process becomes more efficient, because the shell of expanding automated mining probes encounters more and more stars with each passing year. The number of stars per year and the power we receive from mining their gas giants increases with the square of time, and at the cube of the speed we're able to send probes out and return deuterium back.[7] Deuterium mining pays off almost immediately, as the nearest star is four hundred years away at 1 percent *c*, and so in eight hundred years we will start getting back anywhere from one to ten thousand solar-years of energy output from its deuterium, if it has gas giants. Thus, the process already outpaces our Sun's output in the first thousand years, and will only improve after that. By the Year Million, our expanding wall of mining probes ten thousand light-years away should be evaluating around three thousand new stars per year and (assuming life is scarce so ecological plunder is not an issue) we'll be getting back at least seven hundred and fifty thousand years of solar power from these, every year.

Finally, deuterium can be manufactured at great energy cost by making neutrons and combining them with light hydrogen. No one would want to make deuterium this way unless they had unused energy to burn, but that may be the case with energy from lifeless stars that have no gas giants. How can we best condense their energy and get it back to us? Antimatter has been suggested, but storing it presents perhaps insurmountable problems outside our scope here. The much more tractable deuterium is likely to remain the future energy currency of choice. So we might choose to make deuterium inefficiently at stars lacking planets, simply because the energy goes to waste otherwise.

Limitations in Manufacture, and Thus in Time and Space

We have a lot of exploring we can do, if we choose to do it. In a million years, even probes traveling at 1 percent *c* can reach stars ten thousand light-years out. This is a bubble 10 percent of the diameter of our own Milky Way galaxy (which is about half that thick), and 1 percent of the disk's area. If we cover 1 percent of the galaxy's three hundred billion stars, that's some three billion stars. Perhaps the number will be less, since we're not close to the star-rich galactic core, but at least a billion stars to explore in that time and volume seems likely.

So far as we can tell, self-assembling nanotechnology appears physically possible. This means that any physical assembly that's conceivable, stable, and made of atoms can probably be manufactured to order, as many times as one would like. This includes biological organisms, which should be as copyable (and transmittable) as photographs are now [as discussed by Wil McCarthy in Chapter 6—Ed.]. Self-replicating factories should take hold in any solar system, building whatever can be compiled out of ordinary matter, using matter and free energy available locally. These factories will be limited by physical laws, but such limitations (even by the laws we know) are wide.

Alien cultures (if any exist) and certainly our own culture (if it survives) are thus likely to develop general atom-by-atom constructor and duplicator machines. These should be able to remake themselves on the basis of a seed amount of information, wherever time, energy, and suitable matter is available. The seed amount of information for a universal constructor appears not to be large. Ribosomes in living cells come close to being universal constructors of biology as we know it, and they (and the DNA/RNA information that drives them) are fairly simple and small. Doubtless constructors of greater ability are possible and will need more information, but information storage can be much more compact than the 165 daltons per bit that DNA uses in Earth organisms. A packaged fertilized seed or egg is a good metaphor for any Von Neumann–type self-replicating machine.

One of the corollaries of these ideas is that, if we meet space-faring aliens from other stars, they are extremely unlikely to arrive *Star Trek* style, as some kind of "meat in tin cans" traveling between stars in faster-than-light vehicles. Instead, sublight velocities will be used, and matter sent at these relativistic speeds will be kept to a minimum, in order to conserve energy, allow for decent speeds, and maintain the ability to stop at the end of the journey. The problem of slowing matter down at the opposite end of star flight is far worse than the problems of accelerating it at the beginning (which can be done electromagnetically, or by laser). A probe traveling at 1 to 10 percent of the speed of light cannot be stopped directly by a solar sail. Schemes that use giant mirrors to decelerate other sails or mirrors, powered by coherent electromagnetic beams from *another* star, are not convincing. Solar sails work well for star flight only with lasers or masers at both ends. Landing at the "other end" is hard if there's nothing there.

A better idea seems to be Orion-type craft (powered by those mini-nuclear bomb blasts) that possibly can be constructed to travel at 0.01 *c*. However, the material costs required are gigantic. Another alternative for hardy seed probes is "photosphere aerocapture," a one-time pass through a target star's atmosphere, allowing gravitational capture of the craft. This is difficult: only the outer photosphere (visible to us as a star's "surface" and typically about 10 percent Earth sea-level gas density) is suitable for the job, but so thinly-layered that a probe will have to skip through it like a stone off water. Too shallow a dip will give too little energy loss; too deep and the frictional temperatures and *g*s will destroy any object.

Probably only diamonds and carbides, cooled by ablation, will survive a one to ten second photosphere pass. Electronics now survive tens of thousands of *g*s in artillery shells, and the Galileo Jupiter-entry probe reached 15,000° K (far hotter than the photosphere) in a two-minute deceleration from 0.00016 *c*. Both parameters are far from ultimate limits. Photosphere aerocapture requires a minimal perihelion velocity of 0.002 *c* (escape velocity from a Sun-like star at that distance) but allowed excess interstellar energy above this will be set by the toughness of the seed.[8]

Such star-diving probes can be small, because they don't need reaction drives. After slowing to capture velocity, a probe must pass though the target system looking for suitable rocks to impact and then transform into welcome stations. These factories can be used first to build habitats for visitors, and then finally compile the visitors themselves. Think of Sir Arthur C. Clarke's *Rendezvous with Rama*, only on a much smaller scale. Whether the information for doing all this comes with the probe or is sent later by laser is a detail. (Probably the laser will be needed in any case, to send updates for those users of Windows still struggling to eliminate worms and viruses.)

So forget multigenerational ships full of large organisms like humans, or even cryonically preserved corpses awaiting medical resurrection. We shall long since have redefined "human" and found how to make intelligent matter and its mobile representatives, as needed.

We Meet The Aliens: Hey, They Look Just Like Us!

Which is to say, they look like everything, but not particularly like anything. "Humans" exploring another star system will be able to take on any material form (a range of them) needed for the environment to be explored. One might expect space-faring aliens to do the same. Indeed, given the "culturally driven" evolutionary changes that might happen over the millennia needed for star travel (remember the exceedingly fast thought processes that proceed even aboard spacecraft, and certainly near the departure star), we might find that "meeting *ourselves*" at another star is just as odd and exciting as meeting aliens who have evolved somewhere else. After all, once technology allows full control of body design, then designs will be based on the physics of the universe plus input from imagination and art. Probably those results will both converge (on some issues) and diverge (on others), to the point that when it's all over, it might not matter how any given intelligent race started, nor what it looked like originally. By that time we'll all *be* our own artwork and engineering, with no traces left from previous slow evolutionary segments. This is perhaps a little disappointing, but worse news is to come.

Navel Gazing and the Fermi Paradox

You might ask yourself: if we don't expect to meet anything more interesting in other star systems than whatever we ourselves can become in a fraction of the time it takes to go there, then why go all the way out there in the first place?

What's more, we will be able to assess aliens remotely, at light speed, before committing ourselves to the teleportation journey, or instead of bothering.

In 1950, nuclear physicist Enrico Fermi realized that space-faring aliens should be quite common, and many of them should have evolved ahead of us, and some located inside of our light cone. It would thus be quite possible for them to have visited us by now (even without warp drive). So, Fermi famously wondered, *where are they*?

A popular suggested answer, in brief: the average high technology phase of an intelligent civilization, counted from their first radio wave emission to the time when they stop emitting, is quite short—perhaps through technical regression, or destruction of their species. Perhaps, as Carl Sagan surmised, aliens always blow themselves up in nuclear wars shortly after the discovery of radio, and that's why we never hear them.[9]

Today, we see that self-reproducing assembler-type nanotechnology is a far greater danger than *mere* nuclear war; in fact, simple nuclear device manufacture from ordinary uranium-bearing minerals (or even the uranium in seawater), is just one subset of nanotechnological prowess. Possibly high-tech aliens have all long since dissolved each other in assembler wars or accidents, and are now melted into what futurians call "gray goo."

Sagan would never have considered a gentler, more renunciatory alternative, namely, that aliens are able to explore but *choose not to.* Star travel with faster-than-light drive involves paradoxes, and star travel at sublight speeds is *boring.* The faster your thought processes, the longer and more boring a typical trip is. One can slow the rate of thinking, so it feels as if you're moving faster, or even suspend it by turning off consciousness (or by riding on a light beam as information). But what's the point? The civilization you left behind will be roaring ahead, and you'll want to talk to somebody when you get home—unless you're prepared to stay away for good. To our descendants, already trying to squeeze the last yottaflop[10] of performance out of their MBrains while orbiting the fires produced by stripping nearby lifeless solar systems of their deuterium, real star travel with its attendant time lag and low energy availability for thinking on the journey is going to be monstrously un-fun. Especially if it is effectively a one-way proposition.

Consider the alternatives. For a hive mind of huge mental capacity (such as the future human race), any type of virtual reality (VR) experience can be had for the asking, as in a dream but with better control and richer content. Why bother constructing a thing or a scene on an exploratory journey in the "real universe" (waiting out the inertia while the atoms get to where they're supposed to be), when you can have the same experience, or a better one, in virtual reality? As thinking rates more and more outpace atom-moving rates, the quality disparity between the two types of experiences is bound to diverge more and more. Real world exploration is bound to suffer—and if residual curiosity about the real world bothers you, it can be turned off.

The Problem of Motivation

The philosopher Schopenhauer noted, "A man can surely do what he wills to do, but he cannot determine what he wills." In the film *Lawrence of Arabia,* the character of

Lawrence makes a similar observation ("A man can do whatever he wants ... but he can't want what he wants"), then plucks at his flesh and adds: "*This* is the stuff that decides what he wants."

In the future, all this will change once the flesh can be engineered. At present, most of the time, humans are driven by our old monkey/mammal motivations, even if we're not always aware of how much. We want recognition (as a stand-in for better reproductive choices, or a shortcut to it); we want money (ditto); and we want education and entertainment. Soon, all of this will be irrelevant. When our drives are no longer controlled by chemistry, or we can modify any drive, feeling, or desire by act of will, then what will happen to us? What determines our will when we *can* determine what our will will be? It's a conundrum without bottom. Instead of exploring, you can stay home and watch *Seinfeld* (or the future equivalent) with your pals in the Matrix. Except that this fine Matrix won't be a product of evil computers. Instead, it's one *you* control, so you can be your own *Neo* or any other Matrix character, without guilt (which can also be turned off if you like, along with the memory of having had it and why). This is choosing the *red pill* with a vengeance.

Versions of the answer to Fermi's paradox thus involve the aliens all staying home in their orgasmatrons, not caring to go out across those cold light-years simply to probe funny-looking primates like us. And what happens when a brain stimulation fully emulates what certain drugs are already reputed to do: obtain the complete satisfaction of achievement without having done anything to "earn" it. Even the disheartening feedback of nonaccomplishment is in theory controllable. One can always think of things just as important as exploration and being an astronaut. And when you think about them, you can feel just as good as if you *had* been an astronaut.

Perhaps the world of reality will end up populated by beings who for some reason made the choice—or have been coerced—to spend some or all of their time "here" in the world of time and space, instead of in the virtual world that runs much faster and promises ecstasy at command. Once it seemed to me unlikely that 100 percent of minds could be susceptible to any "drug" or "infection" or "fad." But that might be wrong. The world of "reality" in the not-too-distant future might end up as nothing more than the province of the "drug resistant" (VR resistant). Ultimately these people might not be there of their own choosing; the idea of choice might not even apply. Intelligence and sapience can exist without self-awareness, a fact we all know from losing ourselves completely in a good film or a hard math problem. This state can be made permanent. It's a handy one to provide for slaves.

Reality slaves may be kicked out of Heaven, in a sense, and sent back as probes and servants into the external universe by lotus eaters with nothing better to do. This idea resembles Gnosticism; as an explanation of our present reality, it's a bit perverse. These exiled minds (self-aware or not) might perform needed real-world maintenance for the lotus-class and their computronium and power sources.

The Deuterium Mines—You're Guaranteed to Like It, if You "Like" Anything

Perhaps navel gazing really does finally capture all tech-savvy races, and is the answer to the Fermi paradox. That wouldn't be very interesting from our present point of view. But it's an uncaring universe—one that probably allows for all its cruel, heedless aspects to be dialed away (at the cost of living in Disneyland), so long as you can program a few other slave entities to be *happy* to go out and get your deuterium for you, or else not think about it at all, any more than a fish presumably thinks about water. So long as the VR habitués feel entitled enough to make deuterium collection the focus of fulfilled existence for some other reality-stuck types, those slaves will be able to do most things. Except change that one little coding feature that makes them phobic about going into virtual reality themselves.

It's rather hard to imagine resisters of the VR opium voluntarily going back to the Hell of reality, just for the hell of it. Or even voluntarily doing so in small bites because (like the old joke) it feels so good when they stop. Miltonic arguments aside, obviously someone will need to "force" them to the deuterium mines and the solar radiator maintenance. Though as we noted, even the ideas of "force" and "free will" become a bit mixed if the only sense of purpose of a designed creature is the satisfaction of *your* desire. Yet without such meddling, the pleasures of traditional attractions in "reality" (including those of power, sex, child care, etc.) can never compete with unadulterated pleasure itself, available in virtual reality. We organic beings have found something like that already, in crack cocaine and "crystal meth"—for some, these brain modulators simply supersede all evolutionary drives.

Looking for Aliens

If they aren't looking for us, then we can recognize them only by how they make their living, like recognizing the lair of a hidden octopus by the litter of crab and clam shells that surrounds it. Advanced aliens presumably aren't radiating radio, nor are they looking for us—long ago, they've imagined far more entertaining stuff than they'll ever get from watching us or talking to us (given our current slow mental processes, it'd be like conversing with a tree). Probably they're busy entertaining themselves at a zillion times our clock rate.

Can we recognize them from afar? Theoretically, energy can be extracted by an advanced civilization from rotating black holes, but at present we have no good way of figuring out how fast black holes rotate even when we spot them (the closest we've found is sixteen hundred light-years away). If our Sun is not near enough to advanced aliens to represent a handy deuterium source, then they (or their agents) probably aren't even interested in our solar system. Presumably all we'd have to worry about is the theft of a few outer planets—but we can hope they'd choose to talk to us before doing anything so crass as engineering an already-inhabited solar system. Natives who are low tech today will be upset about the loss of their nonsustainable resources tomorrow. If aliens do show up, expressing a diffident interest in our spare gas giants, let us not sell mining rights for a few glass beads.

Any far-off alien presence in our slow reality probably doesn't look special. At home, they probably appear as cool dust clouds cloaking good high-output stars. If the manufactured dust clouds are effective thermodynamically, then we won't see the star, just the local Dyson cloud glowing uniformly from within, at a much lower black body temperature. Interestingly, some nebulae and galactic cores do look a little like this. A "gas" halo glowing at blackbody temperatures of several hundred degrees Kelvin will appear as an unnatural "giant brown dwarf"—a star far too large and energetic to have the low temperature it shows. Unfortunately, we cannot directly assess "luminosities" (energy outputs) or sizes of any but very near stars, so any "brown giants" that exist may fool us by appearing falsely to be ordinary (natural) brown dwarves farther away. Similarly we might hope to intercept communications beams from such immense habitats but, alas, the more efficiently compressed a communication is, the more it looks like natural electromagnetic noise. We can expect superintelligences to compress their inter- and intramind chatter thoroughly indeed.

But there is something else that's hard not to reveal—something our own heroically modified solar system will show by the Year Million. In the neighborhoods of so-called Kardashev Type II civilizations—those powerful enough to use their whole solar output—some planetary systems will simply be missing their gas giants. This alone will be hard to detect, but in these systems, clouds of hydrogen gas could be spreading that are suspiciously free of the deuterium originally present since the Big Bang. Has any astronomer tried looking for missing deuterium in molecular gas clouds near other stars? If we see light hydrogen without any deuterium in a cloud that seems never to have been part of a star, that can mean only one thing. No natural process could do this, as we can see from studying our own solar system. A star didn't eat all that deuterium. Some*body* ate it.

Deep Space in Deep Time

Chapter 4

Life Among the Stars

Lisa Kaltenegger

We are on the brink of an amazing discovery: other planets like ours, all around us. At night, when we look up at the stars, we see a few thousand of the billions of stars in our galaxy alone. More than two hundred of them host giant gas balls orbiting at amazing speed, so close to their stars that some of their atmosphere evaporates, like the material of a comet when it gets too close to the Sun. Today, when a comet passes close to Earth, we head outdoors at night to view its spectacular tail spread across the sky. A hot gas giant's atmosphere, blowing off into space, should be an even more amazing sight. It's a spectacle we cannot yet see, because those extrasolar planets are hidden in the glare of their bright primaries. But we do see their stars pulled very slightly by the gravity of the planets as they orbit around them. Sometimes, too, the planets block starlight that otherwise would reach us, when their orbital plane is *just so* as they pass between us and the star. Then our instruments can detect the planet's silhouette and tell us its size.

Some planets we've already observed probably have winds that would flatten us to the ground were we ever to set foot on whatever surface we find there. Many need only forty-eight to seventy-two hours to orbit around their star—a whole year in a few of our days. Some of those huge hot worlds have more than one sun in their skies and probably numerous moons that we cannot yet detect.

A Million Years in the Future

What we know now, however, is by comparison not even a grain in the vast expanse of sand on a long beach. We are just starting out, taking our first tiny steps toward exploring the worlds around us. It's possible that our planet is the only Earth-like planet that exists, but that seems unlikely. Around 20 percent of the observed stars have huge, hot planets orbiting them—gas balls very unlike our own world, mostly bigger than Jupiter. We are starting to find smaller ones, some down to a few tens of Earth's mass, like the super Earths around Gliese 581, although it is much harder to locate them.[1] Still, finding planets smaller than Jupiter indicates that there should be quite a few worlds as modest as ours out there, possibly orbiting some of the stars we see in our night sky.

The first satellite potentially able to find an Earth-like world has been launched, with a bigger one to follow. In a few years, we will have the first craft in space that can look directly at a planet outside our solar system, collecting its light while blocking out the stars. By staring fixedly at the planet for days, we can gather enough light to split it up in a spectrum—a fingerprint of its atmosphere.

This fingerprint will contain nearly as much information as those found on crime scenes. What gases make up the atmosphere? Is there anything we could breathe? Is there a combination of gases that can only have been produced by life? The first planet-hunting telescopes in space will be small, so only big lines in the fingerprint will be detectable—oxygen, ozone, water, methane, and carbon dioxide. Clouds will muddy the first spectral fingerprints like smeared prints at a crime scene. We will learn how to get around the smears; probably we will need to use bigger space telescopes (like using a bigger and better microscope to make sense of smaller features in a fingerprint).

We are already at the start of this amazing journey. That first little spacecraft looking for small extrasolar planets, a telescope only ten inches in diameter, was launched in December 2006, and has the capacity to find a handful of Earth-like planets throughout its lifetime.[2] A bigger telescope (about three feet in diameter) with the same goal is due to launch in early 2009.[3] Both telescopes will look for planets that block out some of the light of their parent stars.

A few years later, the first small spacecraft flotilla and/or a big optical mirror with a central mask will collect the light of any identified Earth-like world and analyze it, giving us a first real characterization of other pale blue dots among the stars.[4,5,6] These missions will produce the first smudged fingerprint of a new world out there. A million years from now, our descendants will know in detail what kind of life is out there in the galaxy, and perhaps they will be out there as well. But if we are lucky, before our present century is over, the first smudged, low-resolution fingerprint of a fascinating new world will be an object of nostalgia.

The first Earth-size planets we find will probably orbit small, cool stars. In the infrared, where the detectable intensity of star and planet signals depends only on their respective temperatures, the contrast ratio decreases, and therefore it is easier to detect the planet as the star gets cooler compared to a planet like ours. These worlds will have a weird feature: for most of them, only one side always faces their star (like our moon, where the same side always faces Earth), making for a permanent dark and bright side. The region where life like us could evolve on such planets depends on how hot the day side gets and how thick the atmosphere is—and thus how well the temperature is distributed between both halves on the planet. Some planets will have comfy day sides and freezing night sides, some comfy night sides and steaming hot day sides; some will have a ringlike comfort zone, where the day side borders the night side, some a habitable climate on the whole surface. Steam could rise up in the atmosphere, and incredibly strong winds could wipe out any structure on the ground—possibly resulting in flat, streamlined organisms or life that dwells under the surface or never leaves the sea.

Carl Sagan, among others, wrote about life-forms floating in the atmosphere of Jupiter and huge gas planets.[7] We have not found any in our own solar system, and have no idea how such systems could evolve, but we are detecting hot, giant, gas planets so fast that we should have ample targets for a serendipitous search in the future. What we must expect is the unexpected.

First, we must seek what we already know how to find, alert for the key fingerprint that indicates life like ours, because that is what we know how to look for. Searching for other, different life-forms would be fascinating, but requires a lot of trial and error—we wouldn't even know what clues to seek. Designing a huge spacecraft for a trial-and-error mission might not be the most efficient strategy, so we search specifically for Earth-like planets while keeping our eyes open for the unexpected.

Eventually, we will be able to place all of the fascinating worlds we find out there into a gallery of planets (first with crude, and then improved, fingerprints) and try to arrange them in a pattern: evolution of a planet from birth to death, evolution of life from birth to death. Some planets in the gallery will have physical signatures that make no sense to us, because at first we won't have imagined a chemistry and climate that could produce them.

How will we know how to arrange the planets' evolutionary stages? We face the same dilemma as a fly looking at humankind and trying to come up with a theory of human evolution. Do humans evolve from big to small, or the other way around? Do eye and hair color form a pattern that indicates the developmental status of a human being, or are they completely unrelated? The age of the parent star will give us clues about the age of the planet, but still the error bars on stellar age will be wide. This is because we will target very "boring" stars to seek planets in the most stable environments, so that fluctuating radiation will not destabilize possible life-forms and won't produce a huge background noise that interferes with the search for the tiny planetary signal.

Wanted: Fingerprints of Earthlike Planets

Finding spectral fingerprints of planets like Earth will be like looking into the past and future at the same time. It provides a glimpse a few million years in the future or even a few billion years into the past.[8] Not only will it tell us about life on other planets, but it will also clarify some of the big unanswered questions about our own world. How could life ever evolve in the first place? Is basic life common? Is complex life common? Is it all mostly carbon- and water-based? Does it need billions of years to develop, as on our planet, or can it develop in a fraction of the time if the conditions are optimal? Is humankind a product of a "bottleneck" situation in evolution—the inevitable outcome of the evolution of carbon-based life? Or are huge asteroid impacts needed to wipe out competing life-forms like the dinosaurs before high intelligence can develop? Any educated guess today will be outdated ten or fifteen years from now, when planned space missions get a first glimpse of planetary light—and our first answers.

Two and a half thousand years ago, a mere four hundredth of the mysterious span ahead before the Year Million, people first started to ask if we are alone in the universe. Imagine having to wait for thousands of years for an answer. Some got burned at the stake for asking such heretical questions, but no force managed to ban that question from inquiring minds. The curiosity that makes us explore the world around us, seeking novelties that help us to survive, also makes us wonder about other worlds among the thousands of stars we see at night now, and this curiosity will persist a million years from now. Could there be oceans under the ice crust of the frozen worlds? What organisms could produce the weird spectral features on some hot moon around a cool gas giant tens of light-years away? Would a sample of early Earth-like planets ever evolve like us?

What We'll Know First

The torus around a star within which a planet like Earth can have liquid water on its surface is a kind of Goldilocks zone for life and habitability—neither too hot nor too cold, but just right. How many planets actually develop life if the conditions are right? We don't know; anywhere between all of them and just one (us). We do have a notion of where this Goldilocks region—the habitable zone—is for different stars.[9] Approaching too closely to the star evaporates so much water that the greenhouse effect in the atmosphere spirals out of control, causing the rest of the water to evaporate, lost finally to space. Yet moving an Earth-like planet too far away from the star will freeze its atmosphere and make the planet progressively colder.

For planets unlike ours, though, this zone is not well understood. A planet larger than Earth should be able to hold on to its water longer.[10,11] Gas needs a higher velocity to escape from a large planet. Even if all water evaporated into the atmosphere, a massive world could hold on to it, so the habitable zone would extend to regions closer to the star. Near its star, a large oceanic world would take a long time to lose all of the water. That in turn would leave some time for life to develop. These big ocean worlds most likely form far away from their host star and then—for a reason we don't yet know—migrate toward the star until, for another as yet unknown reason, they stop, maybe even in the habitable zone.

Even though these parameters sound vague, we are pretty sure that currently close-in, hot, big gas planets must have formed far out before migrating toward their stars because close to their stars there is simply not enough material to build such huge gas objects.[12] Maybe those we observe are ultimately on their way to spiraling into their host stars, and we see only a snapshot in time. Maybe there is a general principle that drives them to migrate inward.

What happens to small rocky planets when a giant planet in their solar system starts to migrate inward? We hope that the Earth-like planets form late, after the sunward migration of the gas planets, because in simulations it is very difficult to keep their orbits stable once the huge gas planets start to move inward.

In light of all this information, the close-in, hot gas planets we already observe could indicate a system of planets, since if the parent star hosts one planet, it could host many more of different sizes, the small ones still undetected, because from the ground our methods of observation are insufficiently sensitive. Or they could indicate that such stars do not harbor Earth-like planets, if close-in, hot, giant gas planets destabilize their orbits. Finding out which is the case will be our first clue to figuring out how planetary systems form, and how rare or common they are.

Starlight, Star Bright

Once we find these blue dots or rocky worlds among the stars, we will look more closely at the planet's light, from starlight reflected by planets, and infrared light, emitted by their warm surfaces. (Imaging with night vision goggles allows you to see warm bodies moving through the darkness just because they are warmer, and thus brighter than their colder surroundings.)

Searching for life like ours, we first look for water. We

believe that life as we know it needs both water and carbon for building blocks, which can be detected in the atmosphere by looking for carbon dioxide. Earth life produces oxygen/ozone, methane, and nitrous oxide. Water, carbon dioxide, and oxygen/ozone are readily detectable in our own atmosphere, because there is lots of it there and it absorbs part of the outgoing flux. If a planet has no atmosphere, then the spectrum we detect has a smooth curve. By contrast, in an atmosphere with chemicals that absorb energy (for example, in the visible or infrared), parts of that energy making up the overall smooth curve will be missing. Spikes and dips in the spectrum reveal chemicals, useful clues for characterizing a planet even light-years away.

Oxygen/ozone and a reduced gas like methane will not exist together if they are not currently being produced. They react quickly with each other and thus become undetectable as soon as production ceases. The high quantities of oxygen, and thus ozone, on Earth come from living organisms. Methane can be produced by bacteria, or just by release of the interior gases of Earth's mantle. Carbon dioxide and water are not necessarily indicators of life, but we think they are needed for life to evolve.

As a second step, after seeking water and oxygen, we launch a flotilla of bigger telescopes into space to look at small features in the planetary atmospheres, such as nitrous oxide and artificially made gases. On Earth, 99 percent of nitrous oxide is produced by life. Freons, for example, are made only by humans, produced by our refrigerators, and so could potentially indicate take-out dinners on other worlds—assuming other complex life failed to figure out that freons destroy ozone. Deducing the existence of other features, such as sea- versus plant-covered mountains, or understanding whether vegetation exists and could be like ours, requires even bigger telescopes and a fuller understanding of the planet.

First Glimpses

Long before the Year Million, these flotillas of free-flying satellites, each with its own telescope, will allow the first close glimpses of new worlds. Any such early picture will be only a smeared image at very low resolution, with most parts of the picture blending into a solid area. Even so, a ghostly outline of a new world among the stars will be an amazing sight.

Putting a bigger flotilla up in space will sharpen that image, and with it our understanding of what's going on in interstellar space. Does Earth-analog vegetation exist elsewhere; is the changing pattern seen on the planet clouds, dust storms, or continents rotating in and out of view? Are dinosaurs or their equivalent still roaming around, or haven't they yet evolved? Are we looking at a slush ball of bacteria where methanogens (bacteria that produce methane as a waste producer) and cyanobacteria (those that produce oxygen as a waste product) struggle for dominance? (Luckily for us, cyanobacteria won out on our planet.)

That level of information for a number of planets will provide a sizable sample of worlds to explore and understand. What makes a planet habitable? What made our planet habitable? A big rocky planet that always faces its star with the same side, as one half of the Moon is gravitationally locked to face the Earth, could provide a habitable environment on its day or night side. Some planets could be mostly made out of water. These so-called ocean planets could evolve life—what would that life look like if it never left the oceans?[13] Considering Earth's evolution, and how long life stayed in each big development stage (e.g., bacteria, animals, humans), the probability is much higher of finding a slush ball covered in bacteria than it is of finding a civilization on a newly detected world.[14]

How long does a civilization live? How long will *we* survive? All the way to the Year Million? Will we figure out how to deal with wars, climate change, and depleted resources? A million years from now, the big question will not be whether there is life among the stars, but rather if humans and our potential posthuman descendants were smart enough to survive to observe it—or *be* it! Pessimistic views of our future focus on trying to figure out what graveyards of extraterrestrial civilizations could look like. Mostly, if intelligently guided activity subsides, nature claims back any detectable signs of civilization within a few centuries (think of ruins in a jungle) [see Dougal Dixon's essay—Ed.].

Can we distinguish an inhabited planet from one yet to develop life? With huge flotillas of satellites in space observing the new world, we can tell a planet with an evolved civilization like ours from one with dinosaurs or simpler creatures on it if the civilizations vent characteristic and unnatural chemical signatures into the atmosphere. As soon as those early industrial signatures subside, though, we can no longer distinguish the two.

Climate changes of the kind induced by civilizations, such as an increased carbon dioxide budget, can be attributed to intelligent activity only if the history and chemical and climate cycles on that planet are known to very high accuracy. After all, some still question the meaning of those findings even on Earth, where we know the climate and chemical cycles to a detailed extent we will never achieve through remote observations of a distant planet.

Extreme Life-Forms on Earth: Possible Life on Other Planets

Terrestrial organisms have an extraordinary ability to adapt themselves to extreme conditions, including cold, acidic, and hot environments. For hyperthermophilic prokaryotes, the optimum temperature for growth is above 353° K (80°C/175°F). *Pyrolobus fumarii,* an iron-breathing bacterium, has been found to tolerate temperature as high as 394 °K (121°C/250°F), setting the record for the highest temperature known to be compatible with life.[15] Extremophiles can live and adapt to an amazing range of environments, suggesting a huge range of possible scenarios for life on other habitable worlds.

Even so, one key question remains unanswered: can life *begin* in such an environment? It can *adapt* to extreme environments, but can it originate there? If extremophiles do not need mild climate conditions in order to evolve, then we can imagine a first glimpse of the amazing variety of habitable planets among the stars, from frozen planets on the outer rim of their habitable zone, with bacteria living below the ice, to planets with high iron content in their atmospheres. With a red-tinted glare, such an iron-rich world on the inner edge of its habitable zone could have striving *Pyrolobus fumarii* organisms. The possible variety of thriving life-forms on other worlds, based on the known extremophiles on Earth, is vast. Whatever color, form, or shapes they come in, bacteria should be the dominant life-form we will detect on planets like ours.

What about communication signals? That depends on how rare it is for communicating beings to evolve as a primary life-form on such planets, and whether such civilizations actually develop forms of communication we recognize. Right now, we are electrical engineers looking for electrical engineers. What's more, any signal weakens over a certain interstellar distance to undetectable strength. Assuming that the speed of light is the upper limit, hypothetical communication with the closest star system would take years (and that's only if someone is taking your call) [a problem discussed by Amara Angelica in Chapter 10—Ed.].

Still, in a narrowly targeted search, scientists are looking for what we know how to look for: radio and optical signals a similar civilization could knowingly or unknowingly emit. If we regard the age of the Earth, 4.5 billion years, as equal to one calendar year, then our civilization has lasted less than a minute. For two civilizations to listen and emit signals with the same technology, they must be at a similar technological stage, as well as physically close enough to each other that the signal is detectable. Yet that would be an astounding coincidence, unless legacy systems have been set up precisely to attract the interest of newcomers. Of course, a signal from another civilization—if we could decipher its content—would be a fascinating discovery, so this narrowly targeted SETI search remains worth pursuing—even if the chance of success seems very low.

Could a moon, rather than a planet, be habitable? In our own planetary system, Titan's hazes and Europa's ice-covered surface might mask any signal of life-forms detectable from this distance. Other moons in our solar system might harbor life, since they have the requisite building blocks. To detect it remotely will be difficult, but attempts will have been completed long before the Year Million. The first tiny step of finding planets like ours out there will be followed by characterizing them, looking for signatures of life and, once it is found, trying to understand its evolution and fate. At the same time, this process will deepen our understanding of Earth's life and our place in the universe.

Whatever we find—millions of planets like ours with thousands of different life-forms, or none at all—we will learn more about our role in the universe. Are we alone in the unimaginable immensity of the cosmos? If so, then we had better take extreme care of the one planet that somehow managed to evolve life from the abundant building blocks we find everywhere. If we are not, then there is an amazing potpourri of worlds for us to discover and explore.

Let me take you on an imaginary journey among the worlds out there, undiscovered by humanity (but well known to their inhabitants, if there are any).

Vast sky surrounds us. We set out, little by little, to discover our galactic surrounding. First, we chart a path to the rocky planets with habitable conditions. Why go there? For the same reasons that have driven explorers through the centuries: a search for resources, fueled by curiosity and the desire for adventure, and eventually the need for new places to live, as the other writers in this section of the book discuss. Will it be worth going to these places to harvest resources? Space travel—if we don't find a completely new way to travel—will be madly expensive over interstellar distances. The closest star is four light-years away, yet so far we have no indication that it has a planetary system.

But there seems little doubt that we will go. "Earth is the cradle of humanity, but one cannot live in a cradle forever," is the famous aphorism of Russian space pioneer Konstantin E. Tsiolkovsky. He wrote that in 1911, nearly a century ago. Will we set out to live among the stars, to colonize all the habitable planets we find? Will we design spacecraft that, little by little, send groups of humans out to new worlds? It seems inevitable, if we want to survive as a species. Not in a million years, but in a billion years or so, Earth's life will need a new planet to live on, as the Sun brightens and expands lethally into the habitable zone.

Of course, this conclusion assumes our descendants will need an atmosphere, food, and a certain tolerable temperature range to exist. If they do, and if humans survive until then, then Earth will have become uninhabitable. Perhaps we will build our own habitats or reconfigure the solar system into a great system of orbiting habitats, as Robert Bradbury and Steve Harris argue in this volume. But even so, some might chafe at the limitations of a single star. Can we breach the limitation of the speed of light, which seems to restrict serious interstellar travel? [Catherine Asaro, in Chapter 5, suggests that this prospect is not entirely absurd—Ed.] Will we use generation-spanning ships to reach star systems light-years away, even if the original crew must die in transit (unless by then they have learned the secret of extended longevity)? If we cannot break the speed of light, then an optimistic assumption is a maximum speed of about 0.1*c* (or 10 percent of the speed of light). A trip to a planetary system 15 light-years away—very close on a galactic scale—will take a minimum of 150 years to complete, and that's one way. How adventurous or how desperate would people need to be to undertake such a journey? But if a life span is much longer, we won't mind spending some of it on a starship, exploring the fascinating worlds around us.

Maybe we will move out incrementally, establishing a first station at a habitable planet, moon, or modifiable asteroid nearby, then moving on after colonizing that world [as Robin Hanson foresees in Chapter 8—Ed.], spreading in a devouring spherical wave. What if we come across a world already supporting organisms of its own? Can we be responsible enough not to destroy their delicate ecosystems, as we have done so often on our own planet?

Even if we colonize space and move out to new frontiers, we will not be able to outrun the problems we find on Earth. Maybe we can delay their impact, but if we do not solve them here, then they will travel with us from world to world like a curse. Even in the few we can already see, there is an amazing variety of planets out there, huge and gaseous and almost certainly uninhabited as yet. We will continue to push the limits of technology to look for the pale blue dots in the galaxy, and when we find them, we will again be at a place in history where we are exploring unchanged territories. What we will learn by studying those unknown worlds will teach us that our own planet, a tiny pale blue dot against the background of space, is one we had better keep safe because, for now, it is our only place to live. It just might stay that way until the Year Million, and beyond.

Chapter 5

A Luminous Future

Catherine Asaro

The progress of the human race could be described as the history of how we didn't know what we didn't know. In a 1924 lecture, the renowned physicist Max Planck described the misleading advice offered to him in his youth:

> When I began my physical studies [in Munich in 1874] and sought advice from my venerable teacher Philipp Von Jolly . . . he portrayed to me physics as a highly developed, almost fully matured science. . . . Possibly in one or another nook there would perhaps be a dust particle or a small bubble to be examined and classified, but the system as a whole stood there fairly secured, and theoretical physics approached visibly the degree of perfection which, for example, geometry has had already for centuries.[1]

Nobel Prize–winning scientist Albert A. Michelson is known for a similar statement made at the dedication of the Ryerson Physics Laboratory at the University of Chicago in 1894:

> The more important fundamental laws and facts of physical science have all been discovered, and these are now so firmly established that the possibility of their ever being upgraded in consequence of new

> discoveries is exceedingly remote. . . . Future discoveries must be looked for in the sixth place of decimals.[2]

Other notable scientists made similar statements at various times. The general sentiment is summed up concisely in Lord Kelvin's famous comment in 1900 at the British Association for the Advancement of Science: "There is nothing new to be discovered in physics now. All that remains is more and more precise measurement."[3]

So much conviction that science had reached its pinnacle—yet, at that very time, the sciences of quantum mechanics and relativity, theories that revolutionized our understanding of the universe and opened entirely new pathways in human progress, were in their infancy. So much remained to be discovered, despite these confident declarations that the game was all but over.

Such quotes have become a cautionary touchstone for many scientists, a reminder that we should avoid letting complacency about prior accomplishments interfere with future enterprises. In little more than one century, we have leapt forward in our understanding of our universe, and in ways that thinkers of previous eras would have been hard-pressed to envision. Yet one area of science—relativistic physics—has in some ways experienced such premature closure. However, relativistic considerations may profoundly affect the development of the human race and what we become by the Year Million.

Superluminal

Many scientists are convinced it is impossible to travel faster than the speed of light. Certainly, according to the knowledge we currently hold, that appears to be so. Relativistic physics doesn't actually prohibit faster-than-light travel but, rather, travel *at* the speed of light. Such travel also involves paradoxes that must be addressed. If we can't go at the speed of light, then exceeding it certainly presents a problem. But this conclusion assumes our theories of physics are complete. To decide we know everything about the plausibility of superluminal (faster-than-light) flight is probably as shortsighted as Kelvin's and Michelson's complacency in the above quotes. In another century—let alone a million more years—scientists might look back at our notions about relativistic travel and shake their heads at our quaint ideas.

Of course, scientists never really say "never." Various groups have come up with theoretical frameworks that allow faster-than-light travel. The problems with these theories are twofold. Many require the expenditure of immense energy, making such travel prohibitive in real life even if it is theoretically possible. The ideas also often rely on mathematical constructs that have no physical meaning, at least not with the science we currently know.

In the 1980s, intrigued by the idea of finding a mathematically plausible way to describe faster-than-light travel, I published a paper in the *American Journal of Physics* titled "Complex Speeds and Special Relativity."[4] The equations with complex speed can be analyzed in terms of complex variable theory, with some rather pretty results. I will summarize them here; a more complete treatment appears in the article. I should note that this is a theoretical game, not a physical result. I use some equations below for readers interested in the mathematics, but understanding the equations isn't necessary to follow the general ideas. In each case, I've also tried to give a physical description of the processes that doesn't require mathematics to follow.

Off the Beaten Path

Although theorists considered the possibility of superluminal particles as early as the late 1800s, the paper that introduced many of the current ideas is the classic 1962 article by Bilaniuk, Deshpande, and Sudarshan on "meta" relativity.[5] In 1967, Gerald Feinberg gave a quantum field theory of noninteracting superluminal particles and introduced the word *tachyon* for superluminal particles, from the Greek word *tachys* (meaning "swift").[6] Various other researchers have compiled bibliographies and written reviews.[7,8,9,10]

To go from subluminal to superluminal speeds, we expect to pass through the speed of light. But according to special relativity, our mass (and thus energy) as measured by an observer becomes infinite at light speed. In addition, we would exist at the instant we reached it for an infinity of time, and our length would contract to zero in the direction of our travel.[11] This isn't a happy state of affairs for interstellar travel. However, physics doesn't actually rule out the existence of superluminal travel. If we could somehow go at superluminal speeds, then light speed would be still an impassable barrier, but now it would be the lower limit of achievable speeds, instead of the upper.

What would current theories of physics predict for such tachyonic particles? Superluminal objects have many entertaining oddities. Suppose we're in a spaceship going at a speed less than c, where c is the speed of light. Two observers are watching us, Sue in ship S and Jake in ship S' (pronounced "S-prime"). Sue and Jake are traveling in the same direction, so we'll call that direction the x axis. Jake is going at speed v relative to Sue, where $|v|$, the magnitude of v, is less than c. Our speed relative to Sue is u, where $|u|$ is less than c.[12] If we define a unitless speed, $\beta = \frac{v}{c}$, then we can define a quantity called the *relativistic gamma factor*, γ, which is given by the following equation:

$$\gamma = \frac{1}{\sqrt{1-\beta^2}}$$

This innocent-looking factor appears in the relativistic equations for many physical properties, including energy, position, and time intervals, and is thus a major source of the superluminal angst created by the theories of special relativity. When $v = c$, the denominator of γ equals zero (because $\frac{v}{c}$ is now 1, and 1^2 is also 1, so $[1-1] = 0$). So the gamma factor is 1 divided by 0. However, in mathematics, dividing a nonzero number by 0 gives you infinity. That means γ—and therefore various physical properties such as mass and energy—becomes infinite for a ship traveling at light speed.

These problems with γ are easily circumvented by representing speed as a combination of a real *and* an imaginary part. An imaginary number is any multiple of i, where i is the square root of -1:

$$i = \sqrt{-1}$$

A number is *complex* if it has both a real part and an imaginary part. For example, we can define speed as a complex number:

$$v = v_r + iv_i$$

In this equation, v_r and v_i represent real numbers. We've added an imaginary component to the speed by multiplying v_i by i, so the γ factor no longer equals zero at $v_r = c$. This treatment is really a type of physics game, because in our present understanding of the universe, the concept of imaginary speed doesn't have a known correlation to physical phenomena. However, it can provide an enjoyable problem that illustrates how equations lead to physical predictions and thus offer us a means to extrapolate future possibilities.

Imagine you are walking along and you encounter an infinitely high tree blocking your path. You can't go farther on that path because it would take an infinite amount of energy to climb an infinitely high tree. However, suppose you could leave the path and walk around the tree. This is essentially what happens when we add an imaginary component to speed; we're going ***around*** the infinitely high barrier presented at light speed.

Consider one dimension, the direction x. If Sue sees our ships travel a distance Δx (delta-X) in time Δt, then the distance and time intervals recorded by Jake in his ship S' are $\Delta x'$ (delta X prime) and $\Delta t'$, where[13]

$$\Delta x' = \gamma \Delta x \left(1 - \frac{v}{u}\right),$$

$$\Delta t' = \gamma \Delta t \left(1 - \frac{uv}{c^2}\right)$$

Depending on the values of u and v, the distance interval $\Delta x'$ can have the same *or the opposite* sign as Δx. In other words, Jake sees us going either forward or backward (compared to him) depending on whether we are traveling faster or slower, respectively, relative to his ship. However, $|uv|$ must be less than c^2 for any value of u and v, so $\Delta t'$ always has the *same* sign as Δt. Thus, regardless of whether a ship goes in the $+x$ or $-x$ direction, we all agree that every ship is going forward in time.

Now suppose our ships go "around" the light speed barrier and achieve superluminal speeds. I'll assume the relativistic equations continue to work even for speeds greater than light. If we ever do achieve superluminal capability, we will probably find as much new physics to describe what happens at those speeds as, say, quantum mechanics introduced to atomic and molecular physics. However, we can get a feel for what might happen by considering what our present-day theories predict.

The Paradox

So we're zipping around at superluminal speeds. Sue and Jake, our observers, are subluminal, so $|u|$ is greater than $|v|$. The above equations then require that $\Delta x'$ *always* has the same sign as Δx; they can record us going in one direction only. However, it's now possible to have situations in which $|uv|$ is greater than c^2, which means that at certain speeds, $\Delta t'$ will be negative when Δt is positive. In other words, we can go into Sue's past. But if something can travel back in time, the effect of an event can be put before its cause. Sue could see you die before you were born!

Time paradoxes aren't the only problem we encounter in the superluminal universe. If we observe a particle moving "pastward," we measure its energy as negative. The famous equation for energy E is given by

$$E = Mc^2$$

where M is the mass of an object traveling speed v:

$$M = \gamma m_0.$$

In our slower-than-light universe, m_0 refers to the mass of an object when it is at rest relative to the observer, so m_0 is often called the ***rest mass***.[14] Suppose Sue observes Jake speeding up. As his ship approaches light speed, the quantity $\frac{v}{c}$ approaches 1 ($\beta \rightarrow 1$), which means γ approaches infinity. In other words, Sue sees both his mass and energy tending toward an infinite value. This is the infinitely high barrier that prevents Jake from traveling at light speed.

Making Jake's speed complex, however, will make M finite at the speed of light. The resulting superluminal travel features many oddities, but we can resolve some of them using the ***reinterpretation principle***.[5, 6, 7] According to reinterpretation, a particle with negative energy going into the past appears as its anti-particle with positive energy going into the future, traveling from its destination to its point of origin. Suppose particle A emits a tachyon with positive energy that travels until particle B absorbs it. It's a superluminal particle, so some observers could measure it going backward in time, which means they would record its absorption *before* its emission. Reinterpretation suggests we actually see the time-reversed process: particle *B* emits an ***anti-tachyon*** with positive energy that travels into the future until *A* absorbs it. We don't all agree on what events we observe, but the laws of physics remain valid.

Reinterpretation offers a temporal analogy of a more familiar phenomenon. Say we're out driving. Our car is moving slower than Jake's car, but in the same direction. If Sue is on the sidewalk, then when our cars pass her, she sees both going in the same direction. However, Jake sees our car moving backward relative to his, and we see ourselves as stationary relative to our car. We all observe different processes, but they are consistent; everyone agrees about the final result (we reach our destinations), but each person sees the events in a different way.

For tachyons, the temporal rather than spatial direction depends on the observer's frame of reference. Say Jake is on a superluminal ship and Sue is a rocket scientist recording his journey from her home planet. Jake always observes himself as moving forward in time because he's at rest relative to his ship. However, his speed is such that Sue observes him and his ship going into the past. What does this mean?

Sue has stumbled onto one of the famous paradoxes for faster-than-light travel. Could she observe Jake go back in time and, say, cause his parents to die before he is conceived? This type of paradox is often given as a reason superluminal travel can't work. However, it is easily resolved. Jake is at rest inside his ship. He isn't moving at superluminal speeds relative to it, so he observes himself in a timeline that always goes into the future. In analogy with the car example, what he observes must be consistent with what every other observer records. It's impossible for him to cause the death of his parents at a time before his conception; otherwise, that event would *already* have taken place in his own timeline, which of course it hasn't. His life as observed by others must be consistent with his own observations. This doesn't mean he can never appear in his own past, only that he can't change that past after *he* has experienced it.

I should note that for Jake to go into the past relative to Sue, he has to travel *very* fast. When he goes back in time, he will be a long way from where he started. To affect his own past, he would have to turn around and come home, which would change his speed and affect when he arrives. The paradox is more complicated than it looks at first glance, but it does leave the door open for Jake to travel the galaxy and end up in places at times before he set out.

What does Sue actually see if she records Jake going into the past?

Suppose Jake travels to point x_1 and arrives at time t_1, as measured on his ship. At x_1, he changes speed such that Sues observes him traveling pastward. This continues until he reaches point x_2 at time t_2 (as measured in his ship), after which he changes speed so Sue again records him traveling into the future. He then continues to his destination, x_3. Sue sees his ship at x'_2 *before* she sees him at x'_1 (the prime indicates the measurement in Sue's reference frame). However, Jake observes himself going forward in time continuously from x_1 to x_3. He records that t_1 is less than t_2 and Sue records that t'_1 is greater than t'_2

A possible interpretation is as follows: At point x'_2 and time t'_2, Sue sees *two* ships created by pair production, one matter and the other antimatter (this requires the presence of enough mass in the vicinity to ensure conservation laws are satisfied). The matter ship travels to x'_3. However, the antimatter ship follows a reversed trajectory between x'_2 and x'_1; it moves backward compared to what Jake observes when he travels from x_1 to x_2. Meanwhile Sue sees a *third* ship approaching point x'_1, a twin to the matter ship now traveling from x'_2 to x'_3. At point x'_1 and time t'_1, the third ship meets the antimatter ship and the two annihilate each other, producing an amount of energy, mass, and charge equivalent to what was used in creating the antimatter and matter ships at x'_2. Although different observers see dramatically different processes, in theory no physical laws are violated. The final result is the same in both reference frames: the ship arrives at its destination.

Of course, the macroscopic nature of this scenario raises problems. For one thing, that much matter-antimatter annihilation creates a *lot* of energy, which could potentially cause side reactions. In addition, the matter and antimatter ships must come into existence at x'_2 without annihilating. It also makes the rather strange prediction that Sue sees the antimatter ship gaining anti-fuel as it travels backward from point x'_2 to x'_1. She could even see Jake uneating the lunch he had enjoyed earlier! The reader may be able to supply additional oddities regarding the scenario, and think up possible explanations.

Imaginary Existence

Another problem raised by superluminal travel is that relativity predicts superluminal objects have imaginary mass. When $\beta > -1$ is put into the equation for the mass M, we get:

$$M = \frac{\pm i m_0}{\sqrt{\beta^2 - 1}}$$

Similarly, energy, length, and time variables have imaginary values. If a universe with imaginary properties exists, we don't know how to interact with it (at least not yet; given a million years, we ought to figure it out). A way to avoid the imaginary nature of tachyons is to postulate that they have an imaginary rest mass: $m_0 = i\mu$ when β is greater than 1 (where μ is a real number). An imaginary m_0 doesn't contradict known physics because nothing can go slower than light speed in a superluminal universe, so a tachyon can never be at rest. Using the imaginary rest mass gives real values for the relativistic mass and the energy. We can do a similar analysis to investigate for the time and length variables of tachyons.[4]

The relativistic equations predict that at a speed $|v|$ close to that of light (so $|\beta| \approx 1$) a superluminal ship that is slowing down experiences effects similar to a sublight ship that is speeding up. As Jake's ship decelerates, Sue measures an increase in its mass, she records time passing more slowly aboard his ship than at faster speeds (time dilation), and she sees the length of his ship contract as it zips by her. Similarly, an accelerating superluminal ship is like a decelerating sublight ship: mass *decreases*, length *increases*, and time *contracts*.

At speeds just barely above that of light, Jake has a huge mass relative to Sue, his time is almost stopped relative to hers, and his length is almost contracted to zero. The faster he goes, the less she records for his mass, the longer his ship appears, and the faster time passes for him. When the ship's speed equals $\sqrt{2}\,c$, the magnitude of its mass is its rest mass, the length is the same as when it is at rest relative to Sue, and time passes at the same rate for both Jake and Sue.

Nothing in the equations sets an *upper* limit on superluminal speeds. Jake can go as fast as he wants. As his speed goes above $\sqrt{2}\,c$, more delightfully odd things happen; Sue records a *decrease* in the magnitude of his mass relative to his rest mass, an *increase* in the magnitude of his length relative to his length compared to when his ship is sitting on the launchpad, and an *increase* in how fast time passes for him relative to her reference frame. If he accelerates all the way to infinite speed, his mass goes to zero, his length is infinite, and an infinite amount of time passes for him while no time passes for Sue. A strange state of affairs indeed! The term "transcendent" is used for objects with infinite speed.[5]

Luxons are particles that travel at light speed, such as photons. Relativity predicts such particles have a rest mass of zero. At light speed, the relativistic mass is then $M = \frac{0}{0}$. In mathematics, the division of 0 by 0 can actually yield a finite number, much in the way the division of any number by itself equals 1. The quantity $\frac{0}{0}$ isn't always 1, and to determine it exactly usually requires calculus, but the only result we need here is that it can be finite, so it allows the existence of a particle such as a photon traveling at the speed of light.

At subluminal (or superluminal) speeds, such an object ceases to exist, because $M = 0$, so luxons don't seem useful in spanning the sublight and superluminal realms. But perhaps in the future, scientists may be able to use them to form a bridge between the two universes.

Of course, if we did actually learn to travel at superluminal speeds, we would undoubtedly discover other physics that limits our speed to some extent. In fact, most likely we will have to rethink completely how we conceptualize space and time. One advantage to the method I've described here is that superluminal flight could be achieved at relatively low energies. Of course, it also depends on our learning how to give ourselves an imaginary component!

Beautiful Games

To repeat: We know of no physical phenomenon that correlates to an imaginary speed, so the above treatment is really a physics game. It can be fun to postulate physical meanings for the complex quantities. Precedent exists for the imaginary part of a complex function having a physical meaning, as in the damped dispersion equations for the refractive index of light. The real part of the index gives the speed of light in a material, while its imaginary part measures how much the material absorbs light. Perhaps an imaginary speed might also manifest as a physical property we could measure. An analogy between the complex forms of the dispersion and relativity equations, which have similar forms, is given in my paper.[4] The final result suggests some entertaining possibilities, in particular that the speed of light is only one of many singularities. Perhaps if we could go at faster speeds, we might find more infinite trees farther along also blocking the path.[15]

As I discussed in an earlier paper we continually make

leaps in understanding that revolutionize our view of the universe, as witnessed by quantum mechanics and relativity. In the past, such leaps have been accompanied by a reformulation of theories that describe physical phenomena; Newton's equations, for example, gave way to those of Schrödinger and Einstein. The validity of a theory is determined by how well it predicts observed results. Theories of physics are not "absolute truths"; rather, they are models that try to describe the universe to the best of our knowledge. A lack of evidence is not the same as proof of impossibility; theories may turn out to be incomplete when pushed to describe new results previously inaccessible by experiment.

The ability of humans to play mathematical games—a trait shared by few, if any, other species on this planet—often provokes scientific advances, which go hand in hand with new technology. Someday a game or seemingly outrageous theory might lead us to superluminal travel. The history of science is a string of such occurrences. I wouldn't be surprised if we break the light barrier within one or two centuries, possibly with new physics, by learning to circumvent the problems at the speed of light, or by discovering alternative methods of covering interstellar distances that achieve the same result as superluminal travel.

More Space, More Ethical Quandaries

Our species has a history of pushing into new territory. We've reached the accessible horizons and covered the globe, but space remains wide open. Although we may not develop a star-faring society in the immediate future, I've no doubt we will eventually expand into space.

Without superluminal capability, we would need centuries to reach the more distant stars. Our own galaxy is one hundred thousand light-years in diameter, so an expansion through the Milky Way could take many millennia. Reaching other galaxies would take still longer, though by the Year Million, humanity will have had ample time to observe those in our cosmic neighborhood. If we have superluminal travel, everything changes. Our expansion no longer requires millennia, and the prospect of creating a cohesive civilization on an interstellar scale becomes more feasible.

The next few centuries could witness great steps forward in human development as we mold our own evolution. In a million years, we'll be able to change ourselves at will and adapt to any environment. As discussed in other chapters, we will probably lengthen our life span substantially. Already we take for granted lives longer than most of our ancestors'. As biologists continue to unravel the secrets of aging, life spans will continue to lengthen. Someday we may live for hundreds, even thousands of years, which would encourage the settlement of other star systems—assuming transport there and back proves practical and affordable.

If we could travel to the stars with ease, then what sort of "snapshots" might we take of the universe as we move forward to the Year Million? First we will settle the solar system. As we become more adept at dealing with new environments and traveling longer distances, we will spread to nearby systems, then to more distant areas of the galaxy.

Technologies to adapt other worlds to human life will probably be developed within millennia, perhaps even centuries. In popular culture, the idea of molding other worlds has been dubbed *terraforming*. Go far enough along the timeline, and the changes we make won't be confined to worlds. Astronomical engineering may be commonplace as we build, move, or rearrange entire star systems. In a million years, we may be "galaxy forming"—remaking entire galaxies, if we can reach them, for whatever abstruse purposes drive our species by then.

As with any expansion, such prospects bring up ethical considerations. What does it mean to remake a world, a star system, even a galaxy? We see what we will gain, but what will we lose? On Earth, entire cultures have been swept away by heedless human exploration. In my optimistic moments, I see our sensitivity to such questions maturing as our technology matures. Other times, I fear our civilization will drown whatever else we find out there. In a million years, we could remake a galaxy, perhaps a universe, in our own image. But should we?

I doubt the answers will be easy. If humanity reaches the point where we can do galactic engineering, we will have faced these questions a thousand times over. I suspect that as our technology progresses, so will the discussion of our responsibility for how we use that technology. If we keep asking ourselves such questions and keep the dialogue open, we can hope that by the time we gain the ability to reshape galaxies, we will also have the wisdom to find answers.

Group Dynamics

Might humanity someday become a single, great, group mind? It's a fascinating idea with seeds in today's science, an outgrowth of the burgeoning communities forming as the online world expands and our ability develops to combine minds with technology. Although I see the possibility as real, even probable, I'm not convinced that it's inevitable for the entire human race.

As online communities evolve, we're developing an ease of communication unknown at any earlier time. I often hear the term *group mind* bandied about for such interactions. Extrapolated into a future where we can submerge our minds directly into the interactive world, it's possible to see how humanity could become one gestalted mind. If we have the ability to spread our physical selves across the universe, then we might indeed do the same with our minds.

The "group minds" of today represent members of various online communities. Even large communities such as Second Life, though, don't yet constitute a substantial portion of the human population. As the interactive world becomes more integral to our lives, the number of people connected by such technologies will increase, but communities will continue splitting off into specialized interest groups. Some virtual spaces where people interact will grow huge, and there'll be overlap among communities. But will such communities ever represent everyone?

Group consensus is a common theme online; it is the idea that communities form a consensus that represents their members. The evolution of that concept could possibly lead to a consensual group mind for humanity. But wait. With today's communities, do we truly hear a consensus, or the voice of a vocal minority? Consensus or peer pressure? It varies, of course, but peer pressure and vocal minorities affected group dynamics long before the advent of the

Internet and probably will long into the future. And there will always be those who aren't affected by such dynamics. The beauty of the interactive universe is its ability to allow *all* voices to be heard. It works against rather than toward a pressure to conform into one mode of thought.

The value of privacy varies from person to person, depending on cultural background, economic status, and other aspects. Today, some people choose to blog and others don't; in the future, some will choose to open their minds and others won't. We aren't likely to give up that freedom of choice; a shared mind with no privacy brings up the specter of George Orwell's Big Brother in *Nineteen Eighty-Four*. At the same time, like-minded individuals will have the option of joining forces. With an entire galaxy available for human settlement, we are likely to develop uncountable societies and aggregations, some huge and others as small as a pair of friends. Members of some may choose to blend intellects, emotions, even bodies. We may develop thousands, even millions of group minds.

Perhaps in a million years we will evolve beyond the need for privacy. Yet as our ability to alter ourselves advances, so will opportunities to celebrate our differences. Sharing minds could lead to greater acceptance and tolerance, as we learn to understand others by experiencing their own mental states. On my less optimistic days, I wonder if intolerance could also be magnified on an interstellar scale, bigotries based on entirely new forms of altered humanity. The specter looms of entire galaxies pitted against one another. Still, we can hope to avoid such a fate in a universe where humanity has as much room as it wants, the resources of uncounted star systems, enhanced intelligence, the wisdom gathered over extended life spans, and the accumulated knowledge of a million years.

The Eye of the Beholder

As the human species matures, we might wonder if the physical aspects of our existence will become less important. Certainly we are now more concerned with matters of the intellect than our ancestors were a thousand years ago, or at least we have much greater knowledge, and vastly greater access to it. Rather than the focus on our physicality diminishing, however, it has increased as technology has grown more powerful and ubiquitous. Improved health and fitness, greater access to medical care, cosmetic surgery: more options exist now to affect and mold our physical selves. Most adults, given the choice, would appear younger, healthier, more vital, and more desirable. The same options aren't available to everyone, but the further we progress, the easier they are to come by.

The pursuit of mates goes on as always, but it's no longer a foregone conclusion that the reproductive drive results in reproduction. Rather than evolving beyond an interest in sex, humans have begun to decouple the process from the result. As we develop, so probably will our pursuit of intimacy. If anything, the further we progress, the more critical becomes our need for relaxation and pleasure.

The desire to have children is inherent in our species, alongside the drives to eat, sleep, and survive the environment. How we reproduce will change dramatically as the millennia unfold, but the formation of a "family" unit will probably remain ingrained in human culture, whether for the companionship of a mate, a means to raise children, an economic unit, or one of many other arrangements humans use to gather with their loved ones. If we have many planets to settle, I can envision as many different cultures developing as we have ways of thinking about kinship units. Someday perhaps entire worlds will be dedicated to, say, an Amish-like culture that eschews technology. Star-spanning civilizations could be based on any sort of religious or political view, sexual orientation, a desire to share a communal mind, the yearning to wear polka-dotted clothes, any of an unlimited number of possibilities.

When talking about the far future, we assume dramatic advances in genetics, computers, nanotech, and other physical and intellectual changes. I have no doubt we will become just as versatile and knowledgeable about aspects of our humanity that are more difficult to define in this day and age: our emotions. In a million years, we might express love differently, but it's hard for me to imagine calling our species human without love, friendship, or our emotional need to form relationships. People wish to be happy. They wish to be loved. Yet if love survives, won't hatred? What about cruelty, hostility, avarice, and other emotions capable of doing harm? Even those we consider positive can cause pain, and some philosophers have argued that what we consider our darker side might be necessary to what defines us as human.

I would argue that the existence of emotions doesn't require their expression to be destructive. Sometimes futurists see a world where emotions take a backseat to logic, as if the two were different ends of a pole. I challenge that view. I've long thought the traditional separation of emotion and rationality is specious; the two are so intimately intertwined that we can't have one without the other. Emotional intelligence is as critical to our species as analytical intelligence.

Our distant ancestors often seem barbaric when we compare their attitudes and actions to the mores of modern cultures. That doesn't mean we've abandoned emotions. As we've progressed, we have become better able to deal with what they provoke in us. If our understanding of psychology continues to develop, we could dramatically improve the emotional health of our species, bringing our understanding of how emotions and rationality work together as far ahead of our current psychology as our knowledge of physics today is advanced beyond what we knew a hundred years ago.

How does this connect with the expansion of the human race? The mind-set of a species confined to one planet, a few solar systems, or even a slow-moving galactic expansion would surely differ from the attitudes of a species that explodes across the stars. If we tear away the physical boundaries that confine us, then our emotional boundaries change as well. If we can alter our existence on a galactic scale, so should we be able to change our psychology on that scale. It could add a level of maturity to the human species, allowing us to become guardians of our universe rather than of our planet. Still, one could also envision human arrogance on a galactic scale. As an incurable optimist, I choose to believe we will keep our emotional development in step with the wonders we create in the Year Million.

So Where Are They?

Speculation that we might circumvent the light-speed barrier and develop an interstellar civilization cannot avoid considering Fermi's paradox: Where are they, the aliens? If it's possible to create such a superluminal civilization, and the universe is so huge, then why doesn't any species appear to have done it before now?

The question is moot. That we've seen no evidence of any such species doesn't mean they don't exist. The universe is a big place. Really big. At this stage in our development, we can't possibly know what goes on everywhere within it. Possibilities abound. Numerous interstellar civilizations could exist that we don't have the capability to detect. Maybe they are too far away. Maybe they don't want to be detected, or they have evolved beyond where we can find them. Maybe they're so different that we can't interact with them.

Time is another factor. A million years is short compared to the age of the universe. Even if no interstellar civilizations exist now, that doesn't mean they haven't in the past. For all we know, some earlier form of life or a transcendental Creator might have deliberately created this universe by causing or manipulating the Big Bang. And finally, even if no other life in the universe has ever achieved a galactic civilization, well, so what? That's no reason for humanity not to do it. If nothing else, we're a tenacious bunch of biological organisms.

What, then, do I see in a million years? Our descendants spread across the galaxy or even farther in uncountable variations on human life, altered, evolved, and expanded in fascinating ways. My greatest regret in all this is that I won't be around to see what happens. What humans might achieve could be wondrous when the Year Million rolls around.

Chapter 6

Citizens of the Galaxy

Wil McCarthy

Space is big. Really big. It's also hard to reach. The thick atmosphere and deep gravity well of Earth present a real challenge to getting large masses up and away. Fortunately, once you are out there, and if you've got the time, it's possible to move around in space with very little energy.

Newton's inertia might stubbornly resist acceleration, but the vacuum presents no drag on objects that are already moving. What's more, the motion and gravity of celestial bodies can be used to buy acceleration free—remember the Venus-Earth-Earth gravity assist that propelled the Galileo probe to Jupiter? Even better: with proper timing, an Earth satellite in a really high polar orbit—390,000 kilometers high—can hook around the moon every fourteen days as it lopes around the equator. If the Earth and Moon fit on an imaginary plane together, the spacecraft is hopping up and over this plane, and through it, and down and under, and back up through it again, on a perfect schedule. Every time the spacecraft crosses the Earth-Moon plane, the Moon is there giving a little extra tug, so the vehicle is picking up higher and higher velocity until finally it is ejected from the Earth-Moon system altogether. Then you can do the same trick with the Earth and Sun, and step your way out to the larger planets, building up colossal speeds that fire you off like a celestial cannon in any direction.

Given an ordinary rocket engine and a million years, you could easily travel hundreds of light-years from Earth, and by using solar sails and launching lasers or a fancy nuclear engine of some sort [see Chapter 3—Ed.], you could reach any point in the galaxy without ever breaking 10 percent of the speed of light.

The venture is more attractive when you consider you need to make the journey only once. Why? Teleportation. Quantum mechanics might someday allow the equivalent of *Star Trek*'s transporter beam, but even if it doesn't, we can grow a "faxed" copy of you in a remote location, based on a description file transmitted by ordinary radio or laser communication systems.

Think about it: your genome—the complete description of how to build your identical twin—consists of about 35,000 genes, made up from 3.3 billion base pairs, which equals roughly a gigabyte of data. If you store only the differences between your genes and a standardized human genome, and then apply standard data compression techniques, the size of your individual genetic specification shrinks to just a few megabytes. Similarly, since it's possible to estimate your height, weight, physical description, and brain chemistry straight from your genes, all the important attributes of your physical body—including scars, moles, hair style, patterns of gene methylation, etc.—could be stored as differences from the genetic baseline, similar to the way avatars are defined today for virtual worlds like Second Life. A megabyte or two should provide enough detail to fool your own mother. There will be differences, naturally, as there are with the natural clones we call identical twins. Plenty of processes in a growing embryo are stochastic; they follow a prescribed track, but wander around on it.

Then all we have to deal with, in reconstructing you via fax, is your memory. The data storage requirements for this are difficult to estimate, given how little we know about the brain's inner workings, but if we imagine your sense of identity as (a) an *Encyclopedia Britannica* worth of accreted data, opinion, imagery, and sound (five gigabytes), (b) a personal text narrative a thousand times more detailed than a typical historical biography (one gigabyte), (c) a series of three-hour high-definition movies, or thirty-hour low-def movies, showing the highlights of each year of your life (twenty-five gigabytes per year), and (d) a huge relational database that links specific sights and sounds and smells to specific information in the other files, then we could probably get pretty close with two terabytes (that is, two trillion bytes) of storage space. Even if we multiply that by a safety factor of a thousand, we end up with a fairly manageable amount of data, equivalent to about four and a half hours of broadcast time from your satellite TV provider. I'm not suggesting your memories are actually stored in this form, or anything like it, but a future society with a detailed understanding of microneurobiology could almost certainly record, transmit, and reconstruct your memories and personality—your soul, if you will—using this kind of data.

You might find this bald claim hard to believe. Well, think of Mark Twain, or Benjamin Franklin, or Jesus, or any other historical figure you know and love. Chances are you have a pretty good idea what they'd think and say and feel about a wide variety of different subjects. How they'd dress, what they'd sound like, how they'd hold their drinking cup in a fireside chat. Your brain does a pretty good job of reconstructing the person. And yet, all the information you've ever heard about them—their own words, the words of others, a couple of painted portraits—probably amounts to a few megabytes at the very most. So what do you think a hyperintelligent computer could do, one built expressly for the purpose of reconstructing people, with a hundred million times as much information? Even if we allow for some sort of ineffable mystery, even if we accept that something prevents the faxed copy from *being you* in the deepest sense (maybe your unique talents and passions aren't fully captured, or the somewhat stochastic copy expresses them differently), the resulting person would still be more similar to you than are your parents, your children, or even the most identical of twins.

So in a very real sense, you are transmittable and could be beamed to any point in space where someone had bothered to set up a fax receiver. Of course, the strength of even a laser transmission drops off with the square of the distance it has to travel. The TV satellite, from its perch in geosynchronous orbit, has to send its signal only about twenty-six thousand miles (forty-two thousand kilometers) to reach the dish on your roof. To send the same signal to Alpha Centauri would take two hundred million times as much power, or else take two hundred million times as long (about a hundred thousand years of repeated messages for sufficient accuracy) at the same power. Even in a far-off world of near-magical technology, the laws of physics will keep this magnitude of transmission from being a trivial undertaking. The good news, though, is that the blueprints for a whole library of plants, animals, and machines could probably be sent for a fraction of the cost of a single human.

Besides, who is to say that the definition of *human* will remain constant over time? A million years ago there might have been as many as a hundred distinct species of humans scattered around. A million years from now there could be many more than that.

Posthumanism is the science and philosophy of extending human capabilities through a combination of genetic engineering, machine assistance, cybernetic implants, and any other available technology. Think of the Borg on *Star Trek* but in a thousand different varieties, from fairy-tale elves to hard-shelled lobster people to big-headed, big-eyed, gray-skinned *Close Encounters* types—not to mention androids, humanoids, animaloids, floating heads, disembodied brains in jars, hive minds, living buildings, living vehicles, whole landscapes made of pulsing meat. In the extreme case, it might be that each individual is a species unto itself, and the birth of children is impossible without serious technological assistance. Then again, some people won't want to change—will be horrified by the very idea of change—so it's likely there'll be at least a few good old-fashioned Ur-humans running around in the far future as well.

The Energizer Bunny in the Year Million

One thing humans of all varieties will need—have always needed—is energy. Sometimes this is as simple as a campfire or a heating grate to see us through a chilly night.

Sometimes it's more complex: the coal fire that feeds a boiler that runs the turbine that turns the generator that pumps the AC current that feeds the DC charger that ionizes the chemical battery that operates a cell phone. The need might be tiny—just enough to flash an LED for a moment so we know the smoke detector's still working—or it might be enough energy to light a city, or to liberate a ton of aluminum from its prison of bauxite ore. It's hard to say exactly what our descendants will use their energy for, but it's a safe bet they'll use a lot of it, and will be hungry—always insatiably hungry—for more. Some things never change.

As now, people will look to the Sun as the ultimate source. Plants are fine, oil is fine, wind and waves and ocean currents are fine, but ultimately they all get their energy from the Sun. Hydrogen fusion is also fine, but all it's really doing is bringing a small, temporary sun to life down here on Earth. Even fissionable metals like uranium were born in the belly of a supernova. Without starlight, we're nothing.

In a 1959 paper, physicist Freeman Dyson described the ultimate expression of this need: walling off entire stars. His initial concept was simply a swarm of orbiting objects so thick that—like the layers and layers of green in a rain forest canopy—they blocked most of the light passing through them. Today this is referred to as a "Dyson swarm" and its corollary—a solid shell of material completely surrounding the star—is called a "Dyson sphere." Although the light of a star would be blocked by such obstructions, its heat would not. Dyson therefore proposed searching for extraterrestrial civilizations by looking for mysterious points of infrared—of heat without light—out there in the cosmos.

These ideas were directly inspired by the novel *Star Maker* by Olaf Stapledon, who rightly deserves equal credit, but it was Dyson who did the math and concluded that by harnessing the Sun's entire energy output, such a structure would increase humanity's available power supply by a factor of some thirty-three trillion. Wow. I doubt anything like a full Dyson swarm will ever be built, but if even one millionth of the job is completed over the next million years, we'll have thirty-three million times as much power as we did in the year 2000. That's nothing to sneeze at.

Still, by then we might have thirty-three million times as many people as we do now, or each person might, for one reason or another, require thirty-three million times as much power to get them through their daily lives. Energy consumption does always seem to outpace the supply, and with antimatter factories, cryogenic cooling traps, and a steady chatter of outgoing fax broadcasts browning out the power grid as the air conditioners do in our cities today in high summer, the human species (whether singular or plural) will need to look elsewhere for more. We will need, in short, to expand the supply.

Growing Life Elsewhere

But to where? In recent years, we have learned that planets are common around observable stars [see Chapter 4—Ed.], and there are sound biochemical reasons to believe that life might be common as well. However, if the "Rare Earth" speculations of Peter Ward and Donald Brownlee are accurate, *complex* life might be exceedingly rare, leaving Earth-like planets in extremely short supply. Novelists Jack Williamson and (again) Olaf Stapledon are credited with the concept of "terraforming"; that is, deliberately modifying the atmospheres and surfaces of alien planets in order to render them habitable by Earthly life. This is typically envisioned as a centuries-long process. Similarly, various Earth organisms are thought to be capable of surviving in certain extraterrestrial environments, and we can safely expect that with bioengineering and other tools, we can adapt our favorite life-forms for existence on a variety of alien planets, and just fax them there.

But shipping a whole planetary ecosystem by fax could tie up the machine for thousands of years. Why not send a seed instead? In principle, if a mushroom can grow from a microscopic spore, then a gigantic "womb beast" could give birth to a steady stream of other organisms, including more womb beasts. Call it *nanocolonization*: the encoding of an entire Noah's ark into a tiny, easily transportable package the size of a grain of salt. In fact, why wait for the fax gate at all? Just scatter these babies on the cosmic wind, or fire them like machine-gun bullets at distant star systems, so the planets are already habitable by the time we arrive.

And why stop there? While we're at it, we can pack an entire technological infrastructure in there as well: the spore can grow the machine that builds the factory that wires up the entire solar system. The pièce de résistance, of course, is to build the fax gate once the terraforming job is done and switch on its homing beacon, welcoming humans and posthumans alike to an archipelago of fully pretailored worlds.

With such a strategy in place, fresh star systems could be settled in a massively parallel way. The same brave pioneers could send copies to a thousand points of light out there in the cosmos. The complete process, from the arrival of the seed to the establishment of a fully functional human society, might take as little as ten years. A mere decade? What about the centuries-long process of terraforming? Well, with nanocolonization, you don't have to wait for a stable ecology to establish itself layer by layer. Instead, you just rewrite the planet's litho-hydro-atmosphere in a few broad strokes. Even if it takes a thousand years, though, the implications are profound: the spread of humanity (and posthumanity) is limited mainly by the velocity of the spores. This can't be too fast, or they'll vaporize on impact, but as the analysis of Martian meteorites has shown, a planetary encounter at speeds of approximately ten kilometers per second (about 22,000 mph) is not necessarily going to sterilize a sample. Ergo, if the seeds are intentionally designed for impact using advanced future technology, we can probably expect them to handle a couple of additional orders of magnitude on the velocity vector, or roughly 0.3 percent of light speed.

Now, the disk of our galaxy is three thousand light-years thick and one hundred thousand light-years wide, so a million years is only enough to colonize about one two-hundred-and-seventieth of its volume by these methods. However, if we're also sending out 0.1*c* starships in a more limited but more focused way—paid for by the same methods as space tourism today, but on a much larger scale—then we could have a sparse but permanent presence in all but the farthest corners of the galaxy by the Year Million. A profound implication of this expansion is that if there are intelligent aliens out there somewhere, we're likely to encounter them.

The Silence of the Aliens

How likely? In 1960, University of California astronomy professor Frank Drake proposed a method—now known as the Drake Equation—for estimating the number of intelligent, radio-reachable civilizations currently living in our galaxy. By filling in the blanks in his equation with his best guesses, he came up with the number ten. In the entire galaxy. A decade earlier, though, physicist Enrico Fermi had already published a sort of pre-rebuttal now known at the Fermi Paradox: if the galaxy really is hospitable to intelligent life, why hasn't anyone visited yet or tried to make contact with us? Indeed, if a million-year civilization can stretch its tentacles across the entire galaxy, why isn't the evidence of their presence all around us? Cute apologies aside—they use ESP rather than radio, they're busy immersed in Second Life, they are communing with their gods, they *are* gods—the answer is statistically obvious: when all is said and done, and all the variables are known, the Drake Equation will yield an answer significantly less than one. The first civilization to make it to the stars will take possession of the whole shebang, and given the fourteen-gigayear age (or apparent age) of the universe, the chance that someone is already partway through that million-year process right now is barely one in fourteen thousand. I don't like those odds.

This is partly good news for our descendants, because

it means they won't be fighting any interstellar turf wars along the way. Not with aliens, anyhow—they might occasionally slug it out with each other. It's partly bad news, too, because it means we won't be inheriting an alien *Encyclopedia Galactica* that answers all the hard questions for us. There'll be a lot of mistakes and a lot of hard lessons painfully learned. On the other hand, there might be a lot of intelligent species out there who, for whatever reason, never took to the stars or the airwaves, and we might someday end up playing Wise Nurturing Elder Race to them—assuming, of course, that our colony spores are smart enough to avoid accidentally terraforming over them!

Squeezing the Moon

Or rearranging their solar system's mass in interesting ways. Hollow asteroids and dome cities, components of the traditional Dyson swarm, might serve as homes for a great many of our descendants (unless the builders have chosen to migrate into cyberspace, as we'll see below). So, too, might "planettes," artificial bodies resembling planets but with unnaturally dense cores that allow them to be much less massive than regular planets, while still retaining approximately Earth-like gravity and a thick layer of breathable air. Remember *The Little Prince*?

The minimum size for a stable planette depends on a number of factors, especially the temperature and density of the core and atmosphere. The most important factor is that the speed of air molecules has to be much less than the escape velocity of the object, as determined by its surface gravity. The Moon, for example, could be compressed into a permanently habitable planette, with exactly Earth-like gravity, if its diameter were reduced by 60 percent.[1] Arguably, this is much safer than traditional tin-can space colonies, and a more efficient use of mass than locking it up in full-sized planets. By surrounding the old Moon with a thin jacket of antimatter, or some fusionable solid such as lithium deuteride, we could create a planet-sized bomb that, like the implosion-style plutonium bomb dropped on Nagasaki, produces huge pressures uniformly around the entire surface. In the case of plutonium, this increase in density triggers a nuclear chain reaction, but as with the lithostatic pressures that turn coal into diamonds or sandstone into quartzite, it could also permanently increase the density of rocks, by compressing the Moon without allowing it to change shape or break apart. Not a cheap project, admittedly, but it's got to be easier than building a Dyson swarm shell around the Sun.

Another good thing to do with excess planetary mass is to turn it into smart matter. Sand is just sand, right? But if we reorganize it into a mass of quantum-mechanical switches, which I've called *transistronium*—though the more widely accepted name is *computronium*—we can (and do) build computers out of it. Why not do the same thing on a planetary scale? Programmer and theorist Robert Bradbury [see Chapter 7—Ed.] has suggested an interesting twist on the Dyson sphere concept: the Matrioshka Brain, a series of computronium shells surrounding a star like nested Russian dolls, each layer powered by the waste heat of the layer beneath it. This structure probably represents the most powerful computer that's physically possible to build, although it might be capable of other interesting tricks as well.

Most of the properties of different materials—not the mass or density or inertia, but their thermal, optical, electrical, and magnetic responses—are really properties of the electron clouds inside them. Since the 1990s, researchers (myself included) have shown that by moving electrons around in clever ways, we can alter the properties of materials dynamically, in real time. This is known as *programmable matter*, and a solid block of it can reorganize in a vast variety of ways, turning not only into solids that resemble various ceramics and metals, but also into complex devices—a telescope, thermoelectric cooler, a laser cannon with superconducting battery . . . and yes, even computronium.[2] So why not build our Matrioshka shells out of that instead? That way, we're not locked into a particular computing architecture if a better idea comes along later. More importantly, we can create other devices within the shells as needed, including fax transmitters, magnetic anomaly generators, and maybe some transparent windows to let a bit of the sunlight escape from one shell to the next.

Still, replacing Dyson swarms with solid shells of programmable matter has one serious drawback: it uses up all the mass and energy in the solar system without really creating any suitable spaces for people to live. Instead, the solar system's entire population will have to be digitized and uploaded, living out their lives as virtual reality avatars in pockets of computronium. It could be worse; living in a virtual world means you have access to limitless virtual resources—food, drink, sex, sport, space, and even time. There's no reason you couldn't be immortal in there, or at least as immortal as the vast computer itself. Eternity as one long Saturday afternoon.

Running Out of Gas

Ay, there's the rub, because within a few billion years our Sun will run out of hydrogen and swell rather suddenly into a helium-burning red giant. That will end the day rather decisively, and even if our descendants somehow manage to migrate the Matrioshka they live in, making larger, thinner shells out of it, the Sun will eventually shrink into a white dwarf, then a black dwarf, and finally a solid chunk of ultracold iron that emits no energy of any kind. If they don't find another home our distant descendants (well past the Year Million, and even the Year Billion), in their virtual quadrillions, will perish.

So there's the other rub, because once liberated from the dull meat of human brains, the Computronians will be free to expand their minds in the most literal sense, to encompass millions of times more processing power and store billions of times more information. It's a godlike existence that we, in the bony prisons of our tiny skulls, can scarcely imagine. But the number of bits needed to represent the person—to transmit the person somewhere else—expands accordingly. It means the fax gates will never have enough energy to send more than a tiny fraction of the population to the nanofabricated Matrioshka of neighboring stars.

Like all the worst addictions, computronium is a trap, a gilded cage hurtling on a one-way journey to extinction.

But there is a way out, if our descendants are brave enough.

Point: there's no such thing as absolute zero. Quantum mechanics forbids it. No matter how cold things get, the Heisenberg uncertainty principle states that the velocity of a particle (well, actually the product of its position and momentum uncertainties) can never drop below a certain minimum value. Even when the Sun has burned out—even when *all* the stars have burned out—the remaining hunks of dead matter will be pockets of nano-Kelvin warmth. Forever. It's not much, but it's something.

Point: at very cold temperatures, the properties of matter change dramatically. The uncertainty in a particle's position grows larger than the particle itself—larger than the spaces between particles in a frozen solid—so that matter overlaps with itself, becomes wavelike rather than particulate, and slumps its way through phase changes we never learned about in junior-high physics. Superfluids with no viscosity. Superconductors with no electrical resistance. Supersolids in which defects propagate without hindrance; smash one with a hammer and the dent will literally fall out the other side and disappear. Stranger things: Mott insulators, checkerboard matter, crystals of permanent microscopic whirlpools. I scarcely know what to say about these, they're so alien to our experience.

Point: because of its wavelike nature, cold quantum matter can store more information than an equivalent mass of hot Newtonian matter. In fact, all possible states exist inside it at the same time. A googol's googol's googol of parallel universes in every grain of quantum sand.

Point: the human will is insurmountable.

Imagine a future so unimaginably frigid that the laws of physics themselves have condensed out as superfluid rain. The rules are different, the ecology complex. But there are human beings, or the descendants of human beings, frolicking there forever among the cold quantum flowers, forming the fluttering pieces of a Heisenberg mind so vast and unknowable that we might as well call it God. Immortal, unchanging, containing within it everything that ever was or might be. Heaven, in other words. And Hell, Gehenna, Valhalla, and anything else the human mind could ever have conceived.

That's way more than a million years off, but unless human curiosity drops sharply between now and then, it's a safe bet our descendants will know what's coming and, like swimmers contemplating a cold, dark pool, will have begun the process of getting their toes wet.

Chapter 7

Under Construction: Redesigning the Solar System

Robert Bradbury

Predicting the development of civilizations is hard if you focus on precise paths. Standing at the fork in a road, all you can do is choose between one path and another. Which one? The critical steps in science are twofold: first, distinguish between possible and impossible paths, and then select from among the possible paths those that seem most probable, while keeping in mind that a variety of different probable paths might end up at the same place.

Ultimately, such convergent endpoints are determined by the laws of physics. Intelligent life-forms could evolve to highly complex yet unsustainable architectures. As an extreme case, a civilization could seek the true answer to one important question, dedicating all of their resources to that end, and leave behind nothing but a slowly cooling collection of iron atoms with low information content but still have answered the "big question" (at least from its own perspective). Such civilizations might be observed for brief fleeting moments by any with the skill to watch for them. Here we are interested in civilizations lasting for at least a thousand millennia, a million years—and those are a limited subset of the phase space for the existence of civilizations, the state of all possible situations in which a civilization can exist.

To understand where we will be in the Year Million, it is helpful to understand where we are now and what available paths lie ahead.

The Twenty-First Century

One day our posthuman descendants will get to work reconstructing the universe, but at the moment, most of it seems to be composed of dead matter collections: galaxies, stars, planets, asteroids, comets, and interstellar dust. Because the paths these collections follow are dictated by laws of physics that we already grasp, we can predict them confidently, but only so long as they remain "dead." Dead paths are relatively clear. Matter that is too hot cannot form complex chemical molecules. Matter that is too cold remains stuck in the form it had when it cooled. Some parts of the universe, however, retain matter in a liquid state with enough free energy to allow new molecules to form, but not so much that they are immediately destroyed. In these parts of the universe, life can develop.

Life follows a complexification path, wherein complexity is dictated by the environment. Static environments lead to simple life-forms optimal for those conditions. But if the conditions vary enough, life-forms can evolve to take advantage of specific conditions (fish in the water, cacti in the desert, bacteria everywhere). A variety of environmental pressures leads to a variety of adaptations. Even so, life still operates under constraints. For example, there are probably many water-covered planets where the evolution of life is constrained by the predominance of a single type of environment. Complex life-forms develop in the water but are limited by their inability to create the high temperatures required for the manipulation of many of the physical elements. Too much water, too little land, no fire. So, in these worlds, metal compounds of particular usefulness are unlikely to exist. High-temperature rocket fuels burning within combustion chambers is impossible there, so the exploration of space never occurs.

Life-Span Extension Phase

Seen looking back from the future, perhaps it will have gone something like this, if we take the right fork in the road:

During the initial phases of deciphering life blueprints, it was not clear that reading the genome sequence of an organism meant it could also be synthesized. The key transition took life from an observed and revered phenomenon to something that could be cut and pasted—"matter as software"—allowing everyone who chose to create life to do so, where all were gods. Enabled by biotechnology and, later, robust molecular nanotechnology, the natural processes of evolution took us into that era.

Engineering the genetic programs for life began with simple modifications, slightly altered genomes designed to produce antibiotics necessary to ward off flesh-eating bacteria, genomes that produced enzymes to clean clothes. Slowly, an understanding of the limits of natural evolution began. When a natural evolutionary choice occurs, it is very difficult to backtrack. On the ancient Earth, a chemical called ATP was selected as the primary energy exchange unit, and the respiratory pathway was selected to produce the ATP.

But there was a fatal bug in this solution: the respiratory pathway produces free radicals that can damage the DNA containing the code on which almost all life is based.

So the evolution of life on Earth selected a set of paths: DNA to store the blueprints, RNA to translate the blueprints into twenty amino acids that, when strung together, produce the proteins sustaining life. Producing the ATP required by those proteins generates those free radicals that ultimately damage the DNA and kill the creature they have built. Nature evolved more than a hundred proteins specifically to maintain DNA in good shape. Maintaining programs, though, is a tricky business. Maintaining them perfectly stalls evolution, because now there is no variety to select from in the next generation. Yet if they are not maintained sufficiently, then everything gets lost to decay.

So these proteins have maintained our DNA sufficiently to express the gains of prior evolutionary paths but not so completely as to block future paths. The problem, to our chagrin, is that evolution does not select for indefinite longevity so much as for fecundity, survivability, and adaptability. This in turn allows organisms to age and perish, generally losing all of the knowledge they have accumulated during their lifetimes. Once organisms grew complex enough that they could acquire large amounts of information via learning during a lifetime of experience, aging (and death as a result of aging) became a significant problem. People slowly began to understand that because billions of humans die and their accumulated knowledge is lost, these natural processes are clearly flawed and need to be corrected.

The development of computer science provided solutions

for data loss. You can store information perfectly and add error-correction codes that allow for perfect retrieval, even though parts of the storage mechanism have decayed to some extent. Methods were developed to provide genome patches for the essential genes that decay over time. Systems were developed to grow complete organ systems for transplantation, replacing failed or damaged organs. Eventually we developed, as a last resort, the technique of transplanting brains into younger bodies specially grown without brains to preserve individual lives—but the old brain carried with it the seeds of its eventual death. Meanwhile, the nanotechnology path continued to develop, eventually yielding complex nanorobots as small as bacteria. These fantastically tiny devices could be sent into human cells to repair or even replace the broken genomes in existing cells, allowing them to function indefinitely. So aging, and the information loss associated with individual death, ceased.

By the Year Million, this great step forward will have been lost in the mists of prehistory. Natural death will be a curse remembered only by antiquarians.

Asteroid Development Phase

Earth's solar system was lucky. Gravitational disruptions caused by its largest planet during the formation of the solar system left an asteroid belt between Mars and Jupiter, where otherwise an extra planet might have accreted. These numerous small bodies provided a wealth of material that could be converted into platforms to harvest the net output of the Sun, not just the tiny fraction captured by the Earth. This resulted in a rapid conversion of the Earth from a pre–Kardashev Type-I civilization (KT-I: where the total amount of energy the Sun delivers to the Earth is harnessed) to a Kardashev Type-II civilization (KT-II: where the entire energy output of the Sun is harnessed). Calculations demonstrated that this titanic project could be completed remarkably swiftly.

The development methods were fairly standard. Nanorobot delivery modules were sent to specific asteroids. The nanorobots set about converting a three-dimensional structure to an extended two-dimensional object (a solar panel, essentially) optimized for harvesting energy. Normal solar radiation provided ample energy to transform small bodies such as asteroids because their gravity well is minimal, making them easy to break up. Once the conversion was complete, each body had at its disposal a significant fraction of the solar output. There were millions of asteroids, but even a single millionth of the available solar energy supply was still a tremendous quantity (10^{26} x 10^{-6} = 10^{20} watts).

The solar system could equally well have been developed using other methods: Mercury could have been dismantled to form a near-star solar harvester, and Mars a middle-distance solar harvester. But the surfaces of Mercury and Mars are at the bottom of relatively deep gravity wells, so their dismantlement was undertaken only after the asteroids were developed and their full potential as energy harvesters exhausted, as we'll see in a moment. By contrast, in certain other solar systems—chiefly those with large planets orbiting outside the habitable zone—life either could never evolve or, if it did, would stall at a stage before Matrioshka envelopment of its star could occur: that is, a KT-I to KT-II transition. Such systems would be stuck in a natural or seminatural state. A primary goal of civilizations intent on developing or nurturing intelligent life elsewhere is to calculate *where* intelligent life might develop, and to determine when such development might represent a threat, from rogue black holes to berserker civilizations.[1]

Note the assumption here that KT-II cultures will value any and all information vectors. Even if a planet can evolve only the equivalent of squids and whales, then KT-II cultures would prefer to leave it alone. By contrast, they would scavenge any system where liquid water, or even liquid ethane, was never available. There is always something interesting to learn from the twists and turns of the natural universe's experiments. KT-II civilizations, particularly those evolved to the level of MBrains, have to determine whether to pursue a survival path, a grazing path, or a replication path, or some combination. Survival path MBrains would largely orbit outside the chaos of galaxies, only occasionally returning for refueling. Grazing path MBrains would migrate from one dead solar system to another, dust cloud to dust cloud, sucking up inanimate matter. Replication path MBrains would seek out close encounter star systems that would permit them to replicate archived knowledge and experience. These cultures have to predict, or in the worst case arrange, with exquisite accuracy thousands, perhaps even millions of years in advance their close-encounter replication opportunities. The further into the future they try to predict the development vectors of undeveloped star systems, the more their compute time skyrockets. Can they be sure another closer Earth-like civilization will not make the KT-II transition and steal the hot dog off their plate as they reach for it. Much, perhaps nearly all, of their computational capacity is dedicated to galactic history and future path simulations.

Planetary Dismantlement Phase

After our solar system transitioned from a pre–KT-I state to a KT-II state, a primary question arose: how to develop the planets? During the asteroid development phase, the shortage of essential elements quickly became apparent. Carbon, the jack-of-all-trades element with the greatest bonding capabilities, was essential. Aluminum was also critical, being relatively abundant and moderately reflective. Silver and gold were rare but prized due to their high reflectivity. Iron and related elements can be magnetized in order to retain information for extended periods.

When civilizations have made the KT-II transition, the shortage of essential elements relative to the amount of energy at their disposal becomes pressing. Everything is a candidate for ruthless restructuring, no matter how sentimental our attachment might be. Matter, especially matter in the shallowest gravity wells, moves onto the short list for restructuring. The era of *solar systems as software* has begun; the boundary between hardware and software becomes very soft. Nanotechnology allows matter to be transformed easily from one form into another. A computer optimized for sorting can be optimized for searching. A mind upload, as an integrated combination of hardware and software optimized for stock selection, can be transformed into a mind specialized for socializing or acting on the virtual Broadway. Many will find the thought shocking, but if this is the path humanity chooses to go down, then we will tear the solar system into shreds and rebuild these relics into computational platforms of many kinds, hosting the equivalent of a trillion trillion human brains. Each layer or orbiting swarm is designed to extract the maximum useful energy from the Sun and vent its waste heat for the benefit of the next layer out, and on and on, all the way to the cold orbit of Neptune.

How do you break up a whole planet for raw resources? Using immense, thin mirrors to redirect available energy, the Sun's entire output can be focused on asteroids, moons, and eventually planets. But while the incineration of planets is feasible in theory, in practice it runs into some problems. Focusing trillions of watts of energy onto a specific point results in the conversion of the matter at that point into explosive plasma that rapidly disperses. This plasma forms a shield blocking further energy delivery. While the vaporization of small bodies like asteroids is feasible, vaporizing moons and planets proved to be far more problematic. It was not that it could not be done. Every nonstellar body within the solar system could be converted into nanomaterial for subsequent development. The question of when and how to vaporize macrobodies was tied closely to how much matter one wished to radiate, and thus lose, into the universe. Bodies could be moved rapidly to various points within the solar system, but only by expending matter. Organizations capturing those molecules could view them either as returnable, subject to taxation, or as salvage material (matter claimed in intersolar space might be deemed their property).

So the disassembly of moons and planets had to be a precisely engineered operation, rather than a gross application of large amounts of energy. Solar-power systems were first sent into orbit, tuned to absorb the frequencies of light available from the reflectance systems scattered throughout the solar system. They then emitted light at precisely those frequencies most useful to power conversion systems on the ground. Transmitting trillions of watts of power across a solar system was not done without significant attention to production and utilization phases of the project. While the energy transmission aspects were relatively fixed, absorption and utilization aspects tended to be site specific, being based on an idiosyncratic mix of elements available in a particular asteroid, moon, or planet. If you had a lot of cadmium and sulfur, then you could engineer quantum dots to absorb energy from a variety of wavelengths. But if you only had carbon and oxygen at your disposal, then the energy manipulation pathways were significantly limited.

So the disassembly of the moons and planets was a stepwise process that depended heavily on the materials available on-site, as well as any that could be shipped economically across the solar system. But eventually all of the moons, and subsequently most of the planets, were disassembled and distributed through the solar system's elemental market.

The bottom line was that the nanorobots disassembled the asteroids first. This allowed one to make the Sun dark from an external perspective (allowing external civilizations to observe that the KT-I to KT-II transition was complete). Then an extensive amount of time and energy was spent relocating elements within the solar system to just those locations where they would be most useful.[2]

Comet Development Phase

Developing the comets took many years—but not millions of years—since comets are effectively lost in space. The Kuiper Belt had been stripped shortly after the asteroid-development phase for the simple reason that those objects were relatively close and had minimal gravity wells. The Oort cloud took much longer. You could send nanorobot transport ships out of the solar system tasked with the mission of disassembling far-flung comets, but the Sun was then but a faint star on the horizon; the problem was lack of available energy. Even when disassembled, a typical comet in its normal orbit would yield only sufficient collecting area to cook a meager breakfast. Pulling comets apart required nuclear fusion reactors, and the materials they require are in short supply compared to the number of comets. By the time you get to comets, it is clear that you are feeding off the dregs of the solar system. Still, the comets were there, they contained useful material in shallow gravity wells, so they had to be developed.

A mixed development strategy was ultimately used. For the largest and closest comets, fusion reactor–powered disassembly ships were sent out to execute the harvesting. For more distant comets, a huge game of solar-system billiards was played. The precise location of all of the comets was determined. Massive computer simulations were run to determine the slingshot trajectories required to nudge them into the inner solar system. One comet is nudged into swinging by another comet whose deflected path causes it to strike a third comet, which then fragments, the largest piece being on the fast track for the inner solar system, where the comet clean-up stations were awaiting their arrival. The optimization criterion was to return the maximum material while wasting the least—with thousands and thousands of years to get the job done.

Stellization of Jupiter and Saturn

Would anyone really tear the *planets* apart? Many will object, "How barbaric!" But, as history tells us, different cultures have different values. What our descendants will have gained at the end of this ultimate strip-mining exercise is a *computer the size of the solar system*. And inside that computer will be . . . our descendants, rendered in digital form, inhabiting vast virtual realities running perhaps a million times faster than our heritage human brains.

Saturn and Jupiter were problems. They contained a colossal amount of material, but most of it was hydrogen and helium, of little use for making computronium. Their cores did contain iron, essential to an information economy. Hydrogen was discarded; the universe still contained plenty, so it could be imported later. Abundant amounts would also be left over from extending the Sun's life span. Helium left over from the disassembly of Neptune and Uranus and the incoming comets was gradually shifted into Saturn's orbit. Saturn itself was disassembled, so that region became an orbiting cloud of helium. To make the helium easier to manipulate, it had been cooled to a liquid state and packaged in Dewar tanks the size of asteroids. So Saturn's orbit became a collection of highly reflective tank cars until the disassembly process was complete. The metal gained by pulling apart Saturn and the outer planets and comets was then migrated inward to begin the process of tackling the disassembly of Jupiter.

Once that immense task was complete, the solar system appeared as follows: the outer planets, the Kuiper Belt, and most of the comets were gone. A huge number of helium Dewar tanks orbited where Jupiter had once been. A computronium swarm-shell orbited where the asteroid belt had once been. Now it supported and maintained most of what humanity had become: a mixed combination of mind uploads and artificial intelligences. The Earth Monument and Venus Experimental Reality planets orbited untouched, and finally a cloud of hot computronium nodes spun inside the former orbit of Mercury.

Given its limited natural ten-billion-year lifetime, the Sun moved onto the engineering task list. Disassembly of the Sun would allow its conversion from a G-type to an M- or sub–M-type star, now with a lifetime of hundreds of billions of years. With available energy, that task could not be accomplished even in a million years. How fast the Sun should be disassembled depended upon a singular decision: how much of the available solar power should be dedicated to disrupting that primary solar energy source? Extending the existing "living framework" required more energy but also more "metals" from which computronium could be assembled and in which memories could be retained. A new star, a helium-based star (let's call it Saturpiter), was required.

So-called helium stars had been identified by astronomers of that epoch: they contained excessive amounts of helium relative to hydrogen, given the usual ratios in the universe. True helium stars had to have been manufactured by technological civilizations; they consisted primarily of helium in order to drive the production of the massive amounts of carbon and other essential elements required by such civilizations. Indeed the label "star" was questionable, because they were in fact gravitationally assisted implosion-driven thermonuclear breeder reactors. They were derived from the earliest thermonuclear bombs, but on a colossal scale. Layering of various materials promoted specific reactions that would ultimately yield as outputs the desired mix of metals. These were triggered to ignite the reaction cascade. Advanced civilizations didn't have time for classical physics to evolve new elements inside supernovas. Under ideal circumstances, an iron core was introduced into a Saturpiter so that the multilayer fusion processes would yield a self-disassembling microsupernova. In less ideal circumstances, Saturpiters were disassembled using processes similar to those used on the planets and the Sun. So the lifetimes of Saturpiters could vary significantly. Because these breeder reactors and microsupernova processes could be cascaded, elements produced by one fed into the next, which was designed to produce a completely different element mixture, so there could be a necklace of Saturpiter-type stars in the solar system at any given time.

In principle, a Saturpiter necklace from afar should be a glorious sight: a ring of stars of various luminosities slowly blinking into and out of existence. In reality, that was impossible, since each Saturpiter was enshrouded by a vast cloud of matter- and energy-harvesting nanomachinery.

In our own solar system, the helium left over from the disassembly of Saturn, Jupiter, and the outer planets and comets was gradually reassembled into the first Saturpiter.

Additional mass was imported from the Sun to promote optimal Saturpiter structures. Typically helium was extracted from the Sun during its coronal mass ejections and transported to the orbit of Jupiter to await the ignition. Removing matter from a star slows the rate of fusion and retards the evolutionary processes of the star itself. This process of "star lifting" had long ago been envisaged, but the simultaneous processes of stellar lifetime engineering with stellar breeder-reactor creation was a direct consequence of the needs of advanced civilizations. A feedback loop existed whereby energy generated from Saturpiter microsupernovas could be funneled back into the Sun to accelerate its disassembly processes.

Except for use in Saturpiters, the material from cores of the outer planets had largely been distributed throughout the outer solar system for information storage purposes. Complex nano-magnet-based storage units orbited where the outer planets had once been. The fast-thought computronium of the inner solar system contained the bulk of humanity. It operated in spans from nanoseconds to minutes. The outer information stores instead dealt with memories that required hours or days to recover for active thought. If one knew what one wanted to think about, then it was necessary to schedule the data retrieval through the large solar-system routers, since the access queues could become quite lengthy. One could view the solar system as an endless flow of bits from the peripheral data-storage modules to the internal computational modules.

Watching the Heavens as the Stars Go Dark

Astronomy had been popular both as a scientific discipline and a spectator sport from the sixteenth through the twenty-first centuries. Conclusions drawn during this period always assumed that observed conditions would continue unchanged. This was the Copernican principle, an attitude of suitable planetary modesty. While that tended to be valid in a "dead" universe, path prediction was much more difficult in one where actors were entering and exiting the stage. As humanity developed, astronomy attained a new status in this postnanobiotechnological era. History could be observed in process. Billions of large telescopes were produced and focused on the entire sky. The evolution of the universe would become a never-ending show.

From time to time, advanced civilizations would see a star go dark, sometimes quickly, sometimes dimming over years. These were advanced civilizations making the Matrioshka Brain transition: the epochal period during which solar systems become highly engineered arrangements of computronium.

Matrioshka Brains

A "Matrioshka" is a Russian doll set wherein each doll can be opened up to reveal another smaller doll inside it. The "Brain" aspect comes from the fact that a significant fraction of the power output of the Sun is used to support computing activities rather than planetary or stellar disassembly activities.

Computronium was the general term used to describe nanotechnology-based computers—meaning mechanical, "rod logic" computers, molecular (chemical) computers, classical electronic computers, or more advanced superconducting computers. For some applications, quantum computers were used. Given all of the energy available, they could be taken apart and reassembled as necessary. As discussed above, it takes some time for the materials to be shuffled around the solar system and reorganized into the optimized, concentric, orbiting shells that make up a Matrioshka Brain (or MBrain). For example, cadmium and sulfur might be separated from Jupiter's core and shipped to the innermost parts of the solar system for use in complex optical computers.

Matrioshka Brains should not be confused with the mislabeled "Dyson sphere." A *solid* sphere enclosing the Sun would be gravitationally unstable and would fall into the Sun after a time. As originally envisioned by Dyson, the disassembled Jupiter would have been organized into larger, O'Neill-type colonies to hold humanity. MBrains, by contrast, would comprise swarm-like shells made from a tremendous number of co-orbiting solar sail–like structures. Careful planning is required, as the specialized sunlight catchers and reflectors have to be organized to allow continual transport of helium and hydrogen out to Jupiter's orbit and allow through the solar radiant energy they'd capture; such an architecture has the ability to transmit, reflect, or refract any solar energy it receives to any (nonblocked) point in the solar system. Each Mbrain shell was larger and cooler than the one inside it. Usually the outer shells used as its power source the waste heat radiated by the shell inside it, although direct sunlight could be made available by sending requests for direct lighting to inner shells. Because the architecture of many of the shell units was akin to solar sails, they could easily be reoriented to allow various amounts of light through to outer shells. One benefit, in a universe that might contain hostile or greedy aliens, is that this design is maximally economical and safe. Would-be rivals have the greatest difficulty detecting you from outside, since the last of the waste energy leaking out into space is the very far infrared, close to the temperatures of the interstellar dust. It is detectable in principle, but hard to find against the background noise.

History of Humanity

In bygone ages, when the knowledge of humanity could be contained in a single brain, it was not uncommon for humans to migrate across Earth and reestablish themselves elsewhere. They migrated, they found food sources, and they formed tribes adapted to their local environments. But once the laws of physics were understood, once humanity had accumulated a vast store of knowledge, migration was out of the question unless it was imposed by brute force. It would be like the Pilgrims leaving behind their knowledge of gunpowder.

Humans understood, eventually, that it was possible to reshape solar systems in order to store knowledge and sustain life for trillions of years. They also understood that this task wasn't feasible across common interstellar distances.[3] It was impossible to transmit the knowledge of a civilization efficiently across light-years. The information had to be copied directly—as one would copy a CD or a DVD. To replicate itself, the civilization had to be put in what was essentially direct contact with another solar system so that distances were measured in light-minutes, not light-years.

That was difficult, given the distances between star systems. The path of a solar system could be redirected, but that took an enormous amount of energy. Vast amounts of computational time were devoted to computing useful future paths so as to minimize the energy required to copy information. For every technological civilization in the galaxy, a burning question was: When would its natural orbit take it to a location optimal for replication? The choice was whether to consume greater amounts of energy and matter redirecting a solar system to a near-term replication point, at the cost of becoming a lesser civilization, or to wait, accumulating the maximal amount of knowledge and history, and duplicate it during the next chance encounter with one of the nearby solar systems. The choices were highly civilization-dependent. Where were you located in the galaxy? What was your forward path over millions of years? Was it possible to scrounge matter, and therefore the material to store knowledge, from near misses with developing stellar nebulae? Each of these questions had to be answered on a civilization-by-civilization basis.

The Galaxy in the Year Billion and Beyond

As the construction of the Matrioshka Brain within the solar system was taking place, humanity was evolving as well. Progress in neuroscience enabled a detailed understanding of how all the complex programs in the human brain combined to produce a mind. Software became ever more humanlike. Humans, with their multithousand-year life spans, stayed longer in virtual realities. Finally it was difficult to tell if the people one was interacting with in virtual realities were actual humans or software constructs. At that point, it was concluded that artificial intelligence had caught up with natural intelligence.

Virtual-reality interactions led humans to understand the limits of their natural minds. Natural senses lacked the bandwidth to engage completely in complex virtual realities. So direct mind-to-computer interfaces were developed. Over time, people augmented themselves by extending into virtual reality. Eventually minds were running more on computers than in their original biological brains. Sometimes, due to rare accidents, or more frequently, by choice, humans began to discard their old brains. More and more of humanity "lived" as uploaded minds in the computronium scattered throughout the solar system.

Venus had been converted into one giant experiment in manipulated evolution. The massive computronium base of the solar system could, of course, simulate natural evolutionary processes, but it was expensive. Sometimes such computations were simply best done using the atoms and molecules themselves. Much of the carbon dioxide burden, with its enormous pressures and greenhouse effect, had been removed from the atmosphere of Venus. The spin had been adjusted so that its day was now close to twenty-four hours. Significant amounts of water had been imported in the form of redirected comets. Nanorobots constructed thousands of independent islands, each of which could be scheduled for experiments in natural evolutionary paths. Genome engineers could design sets of genomes for various animals, plants, planimals, etc., which were then seeded on an island and allowed to evolve naturally. These experiments illuminated the possible directions evolution might take on other planets scattered throughout the galaxy, which, in turn, guided possible choices for future human replication events.

Earth contained a significant amount of useful material from a computronium standpoint, leading to serious discussion: should the motherworld be dismantled? Finally, in large part for historic and nostalgic motives, it was left intact. Very few people still lived there, in a few, multikilometer-high megalopolises, and some more formed scattered communities similar to those of the Amish. The footprint humanity once pressed upon the planet had largely been removed; the world reverted to its natural state, studied extensively as an experiment in the natural evolutionary paths of planets on a larger scale than the islands of Venus could provide.

Interaction With, and Between, Matrioshka Brains

Communication between an MBrain and a present-day human would be pointless. The gap in computational capacity between an MBrain and a twenty-first-century human is ten million billion times *greater* than the *difference* between a human and a tiny nematode worm (which is a mere billionfold difference)! What's more, the evolution of intelligence might not be a linear process. There is a rather large difference between the intelligence of a human and a chimpanzee or parrot, yet their computational capacities are not separated by more than a few orders of magnitude. Accumulated knowledge (language, history, teaching methods, scientific theories, and data) gives the intelligence of individual humans significant leverage. We can therefore expect that the intelligence gap between an MBrain and an entire human civilization would be significantly greater than can be predicted by looking at the gap in computational capacity alone.

Supposing the transition period between KT-I and KT-II civilizations is short (thousands of years) relative to the lifetime of an MBrain (billions of years), it makes little sense for MBrains to concern themselves with creatures much, much lower than, by comparison, insects are to us. A single MBrain could emulate the entire history of human thought in a few microseconds, mere seconds to compute thousands of thousand-year scenarios. Perhaps they'd take an interest once a civilization progresses to the KT-II or MBrain level, the equivalent of a child who might be rapidly educated. Still, the mass and energy resources available to MBrains are so large that they might observe us quite closely for a very long time from a large distance, silently waiting to see if we will make the transition ourselves to an MBrain level.

But while we might be under observation right now from MBrains elsewhere in the galaxy, they are unlikely to interact with us at our current level. More likely, possible future outcomes of pre–KT-I civilizations, like our own, have been computed in some detail. Don't feel too bad! A single MBrain has the same problem relative to a KT-III civilization that we have with it. But KT-III civilizations, made up of 10^{11} or more MBrains, thinking on a radically different timescale from individual MBrains, would meet limits of their own. Since it is likely that the MBrains of a KT-III civilization would be separated by light-years, the propagation delays between them will become a significant problem. If you can compute an answer to most questions before the answer can be received, then why bother asking?

Or are there *already* ten-million-year-old galactic MBrain Oracles that have utilized their capacities for design, simulation, mass transmutation, and star lifting to construct optimal architectures in order to solve specific types of problems? Since the travel time to ask a question and receive an answer from such an Oracle could easily be ten thousand years, the questions must involve problems that are extremely complex and not easily solved ourselves.

Given the possibility that a galactic MBrain/KT-III civilization has existed for several billion years, there could be a large directory of problems and answers computed and stored by MBrains from preceding times. There should be a large amount of information about galactic history: stellar births and deaths, civilization histories, life-form blueprints, and so on. The galactic knowledge base is potentially huge, but is plagued by long latency times for information retrieval, as well as bandwidth limitations if the volume of information is large. While waiting for the retrieval of answers, MBrains might devote their time to devising complex problems that require millions of years of dedicated computation by a dedicated MBrain or closely linked MBrain cluster.

It is impossible to imagine problems of such magnitude, since even one MBrain has sufficient computational capacity to easily solve problems far beyond our current capabilities. Perhaps one ultimate puzzle might be to find a way to tunnel out of this universe into others—or to reverse the dispersal into futile noise threatened by an ever-expanding, accelerating cosmos. If so, the darkened but richly inhabited heavens of a universe filled with Matrioshka Brains might be the mind's last bulwark against that final fall of night.

The Final Galaxy

The ultimate fate of MBrains is unclear. A tradeoff must be made between active thought and information storage. If a massive amount of material is used to construct helium stars, then it cannot be utilized in information storage. A means of utilizing all of the potential energy available might be developed, allowing most of the mass to be converted to iron. The iron could then be arranged—perhaps utilizing other required elements—in the form of a massive static information store. The last available energy could be utilized in accelerating these information stores in the direction of untapped energy sources as MBrain seeds.

And as this solution to energy and habitat issues is discovered again and again by living species throughout the galaxy, and in the galaxies beyond our own, witnesses will notice (as Arthur C. Clarke once conjectured in a different context) that overhead, without any fuss, the stars are going out.

Over centuries and millennia, their brilliant points of light will fade from view overhead. All will be husbanded and harvested, when the tremendous minds of that future epoch swarm and blend and mutate in the computronium shells of their Matrioshka homeland. They will not be remotely human any longer, those great thoughts circulating inside MBrains in the Year Million. Yet they will be humanity's children.

Then, perhaps, they will turn their attention to reconstructing the rest of the universe.

Chapter 8

The Rapacious Hardscrapple Frontier

Robin Hanson

> A hardscrapple life is one that is tough and absent of luxuries. It refers to the dish scrapple, made from whatever's left of the pig after the ham and sausage are made, the feet pickled, and the snouts soused.
>
> Alix Paultre, 2003 [1]

The future is not the realization of our hopes and dreams, a warning to mend our ways, an adventure to inspire us, nor a romance to touch our hearts. The future is just another place in spacetime. Its residents, like us, find their world mundane and morally ambiguous relative to the heights of fiction and fantasy.

In this chapter we will use evolutionary game theory to outline the cycle of life of our descendants in one million years. What makes such hubristic conjecture viable is that we will (1) make some strong assumptions, (2) describe only a certain subset of our descendants, and (3) describe only certain physical aspects of their lives. I estimate at least a 5 percent chance that this package of assumptions will apply well to at least 5 percent of our descendants. Even if these assumptions are drastically violated, this analysis at least offers a simple and clear baseline for comparing other scenarios.

Our Assumptions

Five hundred million years ago brains appeared, and then doubled in size every thirty million years. About two million years ago, protohumans began to double in number every quarter million years, until ten thousand years ago, when with the advent of farming, our population began to double each millennium. With the Industrial Revolution about two hundred years ago, we began to double our economy every fifteen years. Statistics of prior transitions suggest that within this century we might somehow transition again, and then double (incredible as it might seem) *every two weeks.*

This rapid change is mainly the result of all the techniques, insights, and arrangements we accumulate. Since significant innovations are hard to anticipate, it is hard to forecast such abilities through even a few future doublings. So the only way we can see a *million* years in the future is to assume that, by that point, our descendants have long since run up against *hard* limits, limits even we can now foresee. Yes, such limits would defy our can-do, never-say-never spirit, but the future is not an inspirational story.

The situation we'll consider is a region of spacetime toward the edge, by contrast with the center, of a wave of interstellar colonization. We assume the speed of light is an absolute limit on moving matter and information, and that our descendants still won't have seen any hints of nearby intelligent aliens that might hinder their advance. We proceed as a wave of colonists steadily expanding out from Earth, hindered only by nature and each other.

Within this wave, our descendants would have access

to physical resources such as mass, energy, negentropy, and information, all of which are useful for colonization activities, such as making, maintaining, and transporting artifacts and our descendants themselves. We focus here on these activities, and will have little to say about resources not particularly useful for such activities. We assume that after a million years of searching for better ways to use such resources for such activities, technology no longer improves much, and is almost equally available to everyone.

We also assume that at least one key physical resource is highly concentrated somewhere in space, near relatively stable, identifiable oases such as stars, separated by a desert of near-empty space. And we assume that these oases are limited in two ways. First, each has a limited quantity of quickly usable physical resources. To quickly obtain more resources than a single oasis can offer, one must colonize many. Second, oasis resources have few economies of scale across oases. That is, there are few productivity gains from coordinating the development of resources at neighboring oases.

To colonize an oasis, our descendants will send a seed to the oasis, perhaps ranging in size between a grain of pollen and a battleship—a seed that then grows in its command of resources until it becomes capable of sending a new generation of seeds to other oases. (By analogy, picture a plant species adapted to distributing its seeds on the prevailing winds, slowly filling the oases of a lifeless desert.)

We assume that this seed-to-seed cycle is destructive, in the sense that resources used to support this process cannot be quickly reused for other purposes, and that long-distance interstellar travel is neither too hard nor too easy. That is, over the next million years our descendants will not become extinct, but instead, within a half million years, will be capable of making seeds that can move from the near Earth oasis to oases several light-years away.

We make several more assumptions. Our descendants will *not* make a seed that can reliably and rapidly travel a million light-years in a single vast voyage. Most seeds are destroyed en route, but if not destroyed they probably arrive near their intended destination. All else being equal, seeds fail more often the farther they are sent, and so the fastest way to go a very long distance is to stop frequently to develop oases along the way. There are few benefits from earlier seeds clearing or finding safer paths through the desert. We assume instead that oases can see accurately a long way ahead and are very familiar with the kinds of objects they observe.

While it is possible to make predatory seeds that can travel between oases, such travel is usually too expensive to make invasion of occupied oases cost-effective in terms of resources gained, even for very patient invaders. That is, even moderately developed oases find the resource cost to defend themselves against invasion is much less than the cost to build and deploy seeds with a substantial chance of taking over an oasis at which a seed has had a moderate time to develop. Long-distance weapons to destroy oases and their in situ colonists might, however, be affordable.

How our descendants act in this situation will depend not only on the physical and technical constraints they face, but also on their "personalities"—that is, their strategies, goals, plans, and other behavioral tendencies. Some personalities might be inclined to gleefully rape their environment as fast as possible, while other personalities might seek minimal disturbance and sustainable development. What will be the personality mix?

The answer depends on how strongly our descendants *coordinate* their behavior during a crucial era. The familiar biological world contains only local coordination. Genes can coordinate within a body, and creatures can coordinate within a tribe but, on larger scales, genes, creatures, and tribes compete for scarce resources with little coordination. This uncoordinated competition is what allows biologists to use hard limits and evolutionary selection theories to predict biology's rough outlines. If our descendants prove to be similarly uncoordinated, evolutionary analysis might accurately outline their behavior.

Compared to most creatures large and small, however, we humans today coordinate our actions on an unprecedented scale, via markets and governments. But even we do not have a strong planetary government; most of the dimensions in which we compete are not tightly constrained by the United Nations or by global contracts and treaties. Will our descendants coordinate more than we do?

At the extreme, imagine that a strong stable central government ensured for a million years that colonists spreading out from Earth all had nearly the same standard personality, with each colonist working hard to successfully prevent any wider personality variations in their neighbors, descendants, or future selves. In such a situation, the standard personality might control colonization patterns; if every colonist wanted colonization to be slow, for example, then colonization would be slow.

The crucial era for such coordination starts when competitive interstellar colonization first becomes possible. As long as the oasis near Earth is growing or innovating rapidly, any deviant colonization attempts could be overrun by later, richer, more advanced reprisals. But as central growth and innovation slows, such reprisals would become increasingly difficult.

Why? Because with a common unchanging technology and a speed-of-light limit, the farther one falls behind a rapid colonization frontier, the harder it becomes to influence events on that frontier. For example, longer distance weapons are more easily evaded by unpredictable movements. Because of this circumstance, rapidly advancing frontier colonists could act largely independently of those far enough behind them. Thus, once enough colonists with a wide-enough range of personalities are moving away rapidly enough, central threats and rewards to induce coordination on frontier behavior would no longer be feasible. The competition genie would be out of the bottle.

This scenario, of an uncoordinated colonization wave, assumes that sometime after near Earth growth and innovation slows, enough colonists with behavioral variety will get sufficiently far away from the rest, fast enough, to prevent large-scale coordination from much influencing their behavior. And if that uncoordinated competition lasts long enough, an evolutionary analysis would then apply, allowing us to foresee now the outlines of our distant descendants' way of life. Let's find out what our assumptions imply. (The analysis method that produces these conclusions from these assumptions is given at the end of this chapter.)

Oases in the Desert

It seems hard to say much about what styles of music, art, or fiction, or what languages or operating systems, will be favored by our descendants in the Year Million. It is also hard to say much about the kinds of minds they will have. Distinct minds with their own sense of a coherent self, history, and personality might be far larger or far smaller than our own, perhaps broken into millions of parts that interact with trillions of other parts in unfamiliar ways.

Though these aspects are fuzzy, others are clearer. For example, "local" talk is much cheaper than distant talk. Thus, mental coherence is easier to obtain locally than over broad distances, and consequently more uncertainty exists about distant mental states. That is, even our Year Million descendants still have some relatively local beliefs, goals, and plans that have a privileged influence over local actions. And they have sharper memories of those local—by contrast with distant—activities that have led to current activities. Our descendants also suffer tensions between local and distant beliefs, goals, and plans, resulting in interactions analogous to what we call conflict, persuasion, promises, and deals.

Even more so than today, the spatial distribution of activity is highly uneven because of the highly uneven distribution of useful physical resources such as mass, energy, negentropy, and information. Such resources are highly concentrated at those small bountiful oases separated by vast "deserts" (or we could think of them as small islands separated by vast oceans). These oases might be comets, planets, stars, centers of star clusters, galactic centers, or centers of clumps of as-yet-unknown types of dark matter; we cannot yet discern which oases will be important, or how.

The speed of light limits travel, and interstellar distances are so enormous that reliable travel is prohibitively expensive. Still, our descendants sometimes send seeds to cross the vast desert between oases. While the desert's density is very low on average, its total mass is immense; we can now only dimly imagine the dangers and obstacles such deserts might contain. Perhaps "here be dragons." Over such vast distances and times, seeds can suffer internal failures and collisions with stray desert matter—risks that increase with travel speed and distance.

A seed might contain just a single coherent mind, or information for constructing trillions of minds. It will not be a gigantic starship staffed by a human crew, the parochial *Star Trek* or *Battlestar Galactica* vision. At a minimum, each seed contains information and equipment sufficient to turn raw oasis resources into organized productive capital. Seeds can vary in their speed and capacity to carry cargo and fuel. Seeds can have eyes, ears, and scouts to identify and locate oases, other seeds, and obstacles. They might carry weapons to direct at obstacles, seeds, or oases, or shields to protect themselves from weapons and obstacles. They could also have maneuvering abilities, to evade weapons and approach targets.

Once a seed arrives intact at a virgin oasis, it begins to develop resources at that oasis into more usable capital. Such capital would support a largely self-sufficient local economy, and could take the form of mines, refineries, factories, roads, transports, generators, sensor grids, radiation collectors, heat radiators, signal and power lines, construction and repair units, and command and control centers.

Such capital can be used to create more capital, as well as to increase oasis abilities to observe, talk with, hide from, and defend against outsiders. Even moderately developed oases have enough resources to defend themselves ably against invading seeds intent on displacing them. As noted, defense against long-distance weapons might be harder.

Importantly, the capital of an oasis can also be used to launch a new generation of seeds, the first of which might be the original seed refurbished. Capital would be used to construct the seeds themselves, and perhaps launchers to give them a push, or power and information transmitters to help seeds in transit. Oases might also create observatories and simulators to help refine plans for where seeds should land.

Descendants could ride in on seeds or be received later as signals [as Wil McCarthy speculates in Chapter 7—Ed.]. The rate at which capital can be created from oasis resources eventually slows down as oasis resources are exhausted. Different types of oases have different mixtures of resources and different points at which such resources become exhausted.

The Bleeding Edge

What sorts of lives will our descendants live, empowered by the most advanced technology possible and steadily domesticating the vast resources of the galaxy, as they develop oases and travel between them? Will they build immense mansions with armies of servants to reenact famous battles on the south lawn? Will they isolate planets of primitive life to study their evolution, build vast Jupiter brains or MBrains to calculate more digits of pi, or shape molecular clouds into exquisite sculptures visible a billion light-years away?

Perhaps some settled descendants will do such things, but we are considering a realm near the edge of a wave of colonization hundreds of thousands of years after it began expanding away from Earth. Over this vast time, there has been a continuous race to see who can expand fastest, a race that has given the winner of each stage extra advantages in future stages.

This long competition has not selected a few idle gods using vast powers to indulge arbitrary whims, or solar-system-sized Matrioshka Brains. Instead, frontier life is as hard and tough as the lives of most wild plants and animals nowadays, and of most subsistence-level humans through history. These hard-life descendants, whom we will call "colonists," can still find value, nobility, and happiness in their lives. For them, their tough competitive life is just the "way things are" and have long been.

Let us start by considering a wave of colonization with a locally flat wall or wave-front separating a colonized region from a fresh, uncolonized region. This is the sort of wave-front that might pass through any desert peppered uniformly with similar oases, such as the stars in a galaxy. The seeds all fly in nearly the same direction, away from the wall, and fly faster than the wall itself moves. Behind them, most of the good oases are occupied. Ahead, more and more virgin oases are available. A few lucky seeds find themselves at the very leading edge of the wave, and have the luxury of picking a destination ahead with little fear that a competing seed will arrive first.

At the leading edge, all colonists follow similar strategies—that is, context-dependent behaviors. Colonists prefer similar oases, build similar colonies, and stay similar durations before launching similar numbers of similar seeds that move a similar speed and distance. These strategies are similar because they are close to the fastest possible way to move forward. By this time, recall, advanced technology is largely shared in common, and unchanging. Any slower strategies have, by definition, fallen behind and are no longer represented at the leading edge.

Though hardly noticeable to us, small strategy differences might loom large to our descendants. They might tie group identities to such differences, and obsess about whether variations in success are signs of random luck or of quality distinctions. The status ranking is clear; those closer to the leading edge rushing outward into the cosmos are more likely to have better strategies, and in any case will have more influence over future events.

It is not clear how many seeds an oasis typically sends out. But regardless of whether that number is ten or ten trillion, seeds certainly choose ambitious high-risk strategies so that, on average, only *one* seed survives the journey to colonize a new oasis. That is how plants today replicate, a strategy shaped by the pressures of natural selection. Of the thousands of seeds a plant might distribute, on average only one seed produces a new plant that lives to distribute more seeds.

To tolerate such a high failure rate, leading-edge colonists are relatively tolerant of local risks. Any given seed might fail internally or hit a local obstacle. These risks tend to average out across a wide colonization frontier. Colonists might be less tolerant of correlated risks, such as uncertainty about a common characteristic of a larger region; correlated risks do not so easily average out.

Leading-edge descendants are rapacious and hardscrapple. That is, they devote few resources to luxuries that could instead be used to speed oasis growth or seed travel. Oasis colonies try to grow as fast as possible, neglecting long-term consequences for the oasis after seeds have gone, while seeds travel as fast as they reliably can, neglecting consequences for the desert they pass through. An oasis sends out new seeds only when its local capital growth rate is slowing down, and seeds in transit do not stop until the fraction that fail per distance traveled is increasing.

Interactions between leading-edge oases or seeds are probably minimal. However, if there were very fast and effective long-distance weapons, then seeds and colonies might try harder to hide and to choose unpredictable destinations.

When the wave-front is not flat but round, leading-edge strategies would adjust to prescribe slowing down, staying longer at oases and moving more slowly between them, to let more seeds survive and populate the larger volume ahead.

Leading-edge strategies would also adjust to a foreseeably nonuniform distribution of oases or desert obstacles. For example, outside the galaxy the ease of finding oases might fall by a larger factor than does the seed failure rate between oases. If so, the colonization wave would have to move more slowly outside the galaxy. To compensate, colonists would stay longer at oases and travel in slower seeds. Strategies at the galaxy edge would also slow down, in effect building up resources during years of plenty in anticipation of years of famine ahead. Dwarf galaxies would be prized places to replenish resources.

Falling Behind

On average, colonists near the leading edge tend to slowly fall behind the lucky few who remain there as it rushes outward. How does behavior change as they fall behind? It depends on how intense (that is, how selective) was the colonization competition and race so far.

If competition was only moderate, then behind the leading edge we would see a mix of descendants, some who had suffered unlucky setbacks using pretty fast strategies, together with descendants using not quite as fast strategies. In this case, behavior behind the leading edge would look similar to that on the leading edge, but more random.

In the more likely case, wherein the competition has been more intense, it would select for colonists who adapt their behavior well to falling behind the leading edge. After all, the random factors that make one fall behind now might also help one to catch up later. So the best strategy for those who fall somewhat behind is to seek their best chance to return to the leading edge. Strategies that include giving up too easily would suffer on average and eventually fall so far behind that few seeds would be found anywhere near the leading edge.

In this case, seeds behind the leading edge would be not just risk-tolerant but actually risk-loving about their travel speed. That is, they would happily give up a middle speed for a chance to either go much faster or much slower. Expect them to scour technical or terrain uncertainties for unlikely scenarios that might catapult them back to the leading edge.

Since the first colonists to an area grab the best oases, those arriving later must make do with lower-quality oases. Later arrivals must also suffer a larger chance that they will arrive at an oasis only to find another seed already settled in (and perhaps heavily defended). Seeds can try to compensate for such congestion by attempting to signal their intended destinations, by watching and replanning in flight, or by fighting over newly occupied oases.

To compensate for congestion and lower-quality oases, those that fall behind send out more seeds that travel more slowly and less far and invest more in observing, fighting, and maneuvering. Colonists would stay longer at each oasis to pay for all this, invest more in defense, and consider a wider range of oasis types. The net result is that those behind the leading edge move, on average, more slowly than the leading edge and fall even farther behind.

Congestion is a form of interaction, and it suggests other kinds of interaction behind the leading edge. We have so far considered the "lowest rung" of the local food chain; there might also be analogues of parasites and predators that "feed" on this basic colonization process. Working out the details of such a complex ecology is well beyond the scope of this chapter.

Congestion interaction also suggests the possibility of local coordination. A very intense colonization race might select for extreme local cooperation within local colonization "threads." A thread of cooperation would be a volume in space that follows a group of cooperating colonists and their descendants along a path of colonization, ending with a region of the colonization wave-front surface. Such colonists would cooperate to exclude predators and outsiders from their thread, and coordinate to avoid having their seeds target the same oasis. Threads thus avoid congestion costs except on the thread's surface. Better-performing threads would slowly increase their fraction of the surface of the colonization wave-front.

Staying Behind

Even among colonists who use the fastest possible strategies, a great many will find themselves far behind the leading edge and falling ever farther behind. Such colonists become increasingly desperate to find any long shot, however risky, that might move them closer to the leading edge. They would not accept just growing very slowly at poor oases; there is a limit to how slow an oasis growth rate they will tolerate.

So if all colonization behavior were devoted *entirely* to winning the race for the leading edge, the only things to be found a long way behind the leading edge at most but not all oases would be the wreckage of failed seeds, empty abandoned oasis colonies, and a notable absence of the prime resources. It would be as if a wildfire had passed through the area, converting the easy kindling to ash, but burning itself out while leaving slow-burning tree trunks largely untouched.

It is possible, however, to stay competitive at the leading edge for a long time even if from time to time one diverts a tiny fraction of resources at far-behind oases away from the task of chasing the leading edge. If the process of launching seeds sometimes does not quite exhaust all local capital, then the remaining colonists could give up on ever reaching the leading edge and instead devote themselves to staying put.

Such rare, diverted resources are likely, and would entirely change the character of far-behind regions from dead and empty to lively and full. For example, if a few colonists always remained at every occupied oasis, then in short order all such oases would become well developed. Alternatively, if colonists stayed at only a few rare oases, then once those oases developed enough, they could send out seeds to repopulate the local area. Under this second scenario, it would take only a bit longer for almost all local oases to be well developed.

Because seed travel is so expensive, it would cost far more to successfully conquer even a moderately developed oasis than could be gained from the invasion. There is thus a natural defensive advantage against oasis conquest. Long-distance destructive weapons might perhaps be cheap, but they would probably allow one to destroy an oasis, not to acquire its resources.

Because the first colonists at an oasis could repel invaders indefinitely, the distribution of colonist types in a region long after a colonization wave had passed would be determined mainly by the stay-put choices of the leading-edge colonists who had passed by long before. That is, the fraction of local oases dominated by any given type of colonist would depend on the fraction of their ancestors on the leading edge of colonization, weighted by the fraction of oases at which those ancestors left behind colonists inclined to develop that oasis enough to resist invasion and to send seeds to similarly develop neighboring deserted oases. The earlier a colonist begins to stay put, the more of the local deserted oases it could colonize.

The competition for the leading edge selects against devoting resources to stay-put colonists, though only weakly. So to the extent that such stay-put behavior exists anyway, it is in defiance of that selection. Thus such strategies could be somewhat arbitrary (i.e., determined by hard to anticipate details of the way colonist behavior is encoded and varied). So for a long time after a colonization wave has passed, the activity in a region is determined largely by whatever arbitrary stay-put strategies have been chosen by such diverted colonists.

Eventually local selection effects might privilege certain kinds of stay-put behavior over others, but such effects, too, are well beyond the scope of this chapter. It is interesting to wonder whether the universe we see around us is consistent with such a scenario. That is, could a wave of alien interstellar colonization once have passed this way? Perhaps some combination of powerful weapons and oases in hiding from such weapons could be consistent with the virgin appearance of the universe we see.

Conclusion

The future is just another place in spacetime. We have used an evolutionary game-theory analysis to outline basic physical aspects of the cycle of life toward the edge of a wave of interstellar colonization. I have estimated at least a 5 percent chance that this package of assumptions will apply close enough to at least 5 percent of our descendants in the Year Million. Given how very far away one million years is, that would be doing very well at forecasting.

The future is not the realization of our hopes and dreams, a warning to mend our ways, an adventure to inspire us, nor a romance to touch our hearts. I am not praising this possible future world to encourage you to help make it more likely, nor am I criticizing it to warn you to make it less likely. It is not intended as an allegory of problems or promises for us, our past, or our near future. It is just my best-guess description of another section of spacetime. I can imagine better worlds and worse worlds, so whether I am repelled by or attracted to this world must depend on the other realistic options on the table.

Analysis Method

When different personalities choose different colonization strategies, then those strategies effectively compete to see which of them produces a faster expansion out into the cosmos. Of course, from its own point of view, a personality might be happy to lose such a race, deeming the price of winning to be too high. Nevertheless, as we look farther out into the cosmos, we should tend to see the faster colonization strategies prevailing.

If the descendants and future selves of a personality tend to have similar personalities, then the competition in each new colonization generation should select more for the faster strategies at the leading edge. Also, laggards who arrived after the first colonists to an area would have to make do with leftover resources, the resources the first colonists chose to not use, hide, destroy, or defend. Yes, in recent times humans well behind frontiers have competed effectively with frontier humans, but the scale economies that have made this possible are being assumed away for future oases.

After many generations, these advantages should increasingly select for and emphasize faster personalities on the frontier. Eventually, regions anywhere near the leading edge should be populated almost entirely by variations on, and descendants of, personalities that took the fastest winning colonization strategies. These fastest personalities dominate the "way of life" I describe in this paper.

To study these fastest strategies, I have elsewhere constructed a mathematical game-theoretic model to calculate optimal colonization strategies.[3] In its most general form, the colonization problem is described via five key functions:

> An *oasis* function gives the distribution of oases in spacetime by type.

> A *growth* function says how the resources available to a colony depend on oasis type and the time since a seed landed at that oasis.

> A *cost* function says how the resources needed to make and launch a seed vary with features.

> A *mortality* function gives the rate at which seeds fail in transit, depending on the local oasis density as well as seed speed, hardness, and distance from source.
>
> A *targeting* function says the chance a seed successfully gains and holds an oasis of a given type, given the local fraction of occupied oases, the rate of change of that fraction, and the seed's speed and targeting ability.

These functions are usually described abstractly in terms of the signs of various derivatives.

To analyze the optimal strategy in this game, I consider two functions across spacetime: a *density* function, giving the fraction of oases of each type that are occupied, and a *value* function, giving the value of having a new colony there. There are two key consistency equations.

First, a *flow* equation connects the oasis occupancy rate to the rate of arrival of seeds with various features.

Second, a *strategy* equation sets the value of a seed starting to colonize an oasis equal to the best possible sum of the expected value of seeds later launched from that oasis that end up successfully colonizing new oases one generation later. By "best possible" we mean considering all possible strategies for how long to wait at an oasis until sending out how many seeds of what type, how far and how fast. We consider colonist actions that are close to these best possible actions.

As usual in such analysis, to get concrete answers we must make a variety of "technical" assumptions of varying plausibility. While the intent of such assumptions is to simplify the analysis without overly compromising realism, this intent is not always realized. Also as usual, we try to see how far we can get by making relatively weak assumptions, and then we add stronger assumptions as needed to get clearer pictures.

For example, my math analysis has so far always assumed that seeds travel many oasis spacings, that there is only one kind of oasis resource, and that oases are constantly and uniformly distributed in space. Sometimes this math assumes a flat colonization wave-front.

Surely the main question in many readers' minds is: how reasonable is our basic evolutionary modeling approach? This depends on how reasonable are its key modeling assumptions, of "enough" (1) stability of personalities across generations, (2) generations of colonization, and (3) variations of personalities. Whether there is enough personality variation depends in part on the level at which stable behavior tendencies are encoded.

For example, it might be that the stable tendencies are inclinations to build seeds a certain time period after landing at an oasis, or to make seeds of a given speed and hardness that travel a given distance, but with little tendency to *adapt* these behaviors to changing environmental features such as local colonization congestion or the types and density of local oases. If so, an evolutionary analysis might do well at describing average behavior, but poorly at describing how such choices *vary* with such environmental features.

On the other hand, there might be some colonist personalities that want to win at any cost the race to dominate the leading edge. Such personalities would calculate in far more detail than we have done the best behavior to achieve this goal, assuming other independent competitors had a similar aim. With enough initial colonists like this, or with just one or two of them and enough generations of competition, optimal behavior models might do very well at describing behavior near the leading edge.

The Mind/Body in Year Million

Chapter 9

Do You Want to Live Forever?

Pamela Sargent and Anne Corwin

> Every orchid or rose or lizard or snake is the work of a dedicated and skilled breeder. There are thousands of people, amateurs and professionals, who devote their lives to this business. Now imagine what will happen when the tools of genetic engineering become accessible to these people. There will be do-it-yourself kits for gardeners who will use genetic engineering to breed new varieties of roses and orchids. Also kits for lovers of pigeons and parrots and lizards and snakes to breed new varieties of pets. Breeders of dogs and cats will have their kits too.
>
> —Freeman Dyson, "Our Biotech Future"

It's a cliché (and also true) that the future is unknowable. What isn't so obvious is that the past might as well be unknowable, as apparently it already is for large numbers of people, perhaps even the majority of human beings. In traditional societies, the present tends to resemble both the past and the future, leaving little reason to distinguish either from the present.

But a limited sense of time is not confined to traditional societies. Many members of postindustrial civilization have at their command the most sophisticated technology humankind has ever produced, yet even for these people, the past remains as unknown as any future. At best, it's a colorful diorama, glimpsed via the History Channel and the like, providing both escape and maybe a few overly simplified parallels connecting past events to our own time. "Most of the undergraduates and graduate students whom I and others encounter have little or no knowledge of the history of American art, literature, music, or the popular culture of previous generations," historian Richard Pells writes.[1]

In fact, many people seem to lack historical perspective altogether. Anything prior to their birth might as well have happened in the distant past; ancient Egypt and World War I Europe would seem about equally distant and unknown. The future, if it's thought about at all, is imagined to be either a much better version of the present with a few exotic details thrown in or a much worse, dystopian one. This lack of historical perspective probably explains why sizable numbers of people can easily believe that the universe sprang into existence only six thousand years ago, and why, to such folk, an apocalypse foretold by biblical prophecy seems as likely a future as any other.

However, this bubble of historical obliviousness, whose boundaries are roughly that of a human life span, makes it more likely, rather than less, that people in rapidly changing technological societies will accept advances in biological technology that are genuinely radical—even some that amount to what their forebears would have considered blasphemy. For all the talk about how certain biological facts are constants (birth and death being the most obvious), there's a much wider acceptance these days of contraceptive measures, even among very conservative sorts, than many people of our grandparents' time would have thought possible. It wasn't until around 1965 that married heterosexuals in the United States could legally employ contraceptive measures.[2] The idea of expectant couples routinely making use of genetic screening and testing methods would have struck our recent ancestors as still more outrageous; and as far as they were concerned, the less said about gay and lesbian couples, the better.

As for death, one of us is just old enough to recall when people at funerals would mutter about someone who had dropped dead in his late sixties, "Well, he lived a good long life." These days, most in the West would not feel so sanguine about a friend or relative who died at that age. To anyone younger than forty, procedures such as heart-bypass surgery and artificial insemination probably seem entirely uncontroversial, as if they have been around forever. Historical amnesia can be the ally of innovation, at least in gaining acceptance of various innovations, however startling they may be to an older generation.

Repairing the Body

Whatever we accomplish in biology during this century (and it's probably useless to make any specific, detailed predictions, as events are moving much too rapidly for that), we are likely to learn much more about biological processes as *systems*, as organizational patterns. The prospect of patchwork repairs—of surgery to repair this or that worn-out body part—is probably not the way human lives will be prolonged. Interventions are more likely to consist of dietary changes, pharmaceuticals, cellular repairs through nanotechnology, implants of various kinds, and other measures treating the human subject as a system and as part of a larger system.

It's also plausible that more sophisticated enhancements might induce people to modify their own human forms in ways that will make them seem alien to us. Anyone familiar with body piercing, tattoo art, and other such increasingly commonplace modifications (which were also traditional decorative alterations in many indigenous cultures) won't think this prospect unlikely. The basic human form, in various "optimized" versions of the genome, might persist, so that in the Year Million our descendants will still be recognizably human, even if the gap separating us from them will be at least as wide as that between us and our prehistoric ancestors. Or, more likely, the human population will diverge into various subspecies. Some people will favor implants, mind-machine interfaces, and uploading their experiences into computronium substrates, while others will stay firmly entrenched in biological wetware.

Although it is impossible to predict with any accuracy what a far-future society of practically immortal sentients will actually look like, by basing our projections on the diversity of present-day attitudes toward modification we can imagine different factions of humans or posthumans. There will be advantages and disadvantages associated with such technological augmentation as cyborg-style implants, no less than with genetic/somatic manipulation (such as altering the expression of one's own genes to produce greater sensory acuity or athletic ability). Different people are likely to follow their own personal scales and preferences in weighing the costs, risks, and potential benefits of various enhancements for themselves and their children. Regardless of the nature of individual modifications, the increasing range of possible modifications is likely to stress the limits of the status quo considerably. Any society that cannot adapt to a more diverse and extreme spectrum of morphologies and abilities will either be torn down and remade, or drawn into chaos and conflict.

Assume that we avoid the worst catastrophes of global climate change, or at least assume that those who survive them manage to learn enough about the interactions of climate and the environment to overcome disaster. Assume also that we achieve control over the human genome. There's no reason, under these less-than-apocalyptic conditions, to think that life span won't experience a manifold increase, and thus no reason to fear the fate of the mythical Tithonus. In that Greek myth, Tithonus was a mortal who fell in love with the goddess Eos. In a bid to keep her lover alive, Eos begged Zeus to grant Tithonus immortality, but Zeus granted the request only in the literal sense: Tithonus could not die, but he continued to age, becoming ever older and sicker. Many people imagine a similar doleful consequence to further life extension among a growing population of increasingly frail elderly. This cruel scenario is neither the goal nor the likely outcome of new, innovative efforts to increase life span, since health is the primary goal of all medicine. What is sought is a method to prolong the years of health and youthful fitness, compressing and postponing the predeath period of extreme frailty ever further into the distant future.

To accomplish this goal of increasing health span (rather than just life span), we need to determine the precise cellular mechanisms of aging and intervene as necessary to fix the bodily damage that occurs as a result of these mechanisms. One promising route is to look at aging from an "engineering" standpoint, examining carefully the major mechanisms through which the simple passage of time leads to increasing organic and cellular damage, and then to propose practical solutions that address specific categories of damage. This approach, which seeks to heal or prevent damage before it creates serious health problems, differs markedly from traditional geriatrics. Today's geriatric care focuses mainly on mitigating symptoms of age-related disease *after* damage has already progressed into pathology. It differs, too, from the gerontological approach, which focuses on studying the aging process at a detailed level. Engineering focuses on finding solutions rather than just explaining how something works.

The Path to Year Million

If we're to consider drastic life-extension as more than a wistful fantasy, we need to start with a road map derived from today's knowledge, just as any discussion of Matrioshka Brains or galactic exploration must start with today's physics and limited technology [as in Chapters 6, 7, and 8—Ed.]. Fortunately, at least some of the groundwork has already been laid. Laboratory efforts are now being funded and directed toward solving the aging puzzle. While humans have long dreamed of a fountain of youth, we have only recently begun to move beyond mere storytelling and toward tangible progress in longevity medicine. Anticipated health care needs of the aging Baby Boomer population, along with the tenuous state of the future of social-security programs, especially in the United States, are strong motivating factors for scientists and policymakers.

This new motivation, combined with technological advances in areas like computing (which enabled completion of the Human Genome Project ahead of schedule), makes it all the more likely that life spans will start increasing sooner than we could have anticipated a decade ago. So it makes sense for people alive today to consider the implications of vastly longer, healthier lives.

In the first decade of the twenty-first century, the most ambitious proposal for an engineering approach to the mitigation of aging damage is the Strategies for Engineered Negligible Senescence (SENS), first outlined by biogerontologist Aubrey de Grey. SENS is not a prescription for curing aging per se, but rather an attempt to break down the task of addressing aging damage into manageable, logical chunks. In looking at the various differences between old and young bodies, and examining the experimental data gleaned so far from biological and biogerontological research, seven categories of aging damage were identified:

Cell loss and cell atrophy
Junk outside cells
Crosslinks, or protein shackles, outside cells
Death-resistant cells (that is, corrupted cells that we want to die off, but that don't)
Mitochondrial mutations
Junk inside cells

Nuclear mutations and epimutations (which lead to cancer).[3]

In addition to identifying these seven categories of aging damage, SENS also suggests mitigation methods for each damage type, based on existing experimental and theoretical methods. For instance, advances in stem-cell therapy could lead to ways to replenish those cells we need to keep us healthy but that tend to atrophy and die as we age. Analysis of soil microbes and the enzymes they produce could lead to new drug treatments for "storage diseases" like atherosclerosis and Alzheimer's; a research program (Lyso-SENS) at Arizona State University's Biodesign Institute is devoted to exploring this possibility.[4] Mutations in mitochondria—small organelles floating by the hundreds inside each cell's cytoplasm (but outside its nucleus) that convert organic matter into usable chemical energy—might be rendered harmless if their gene expression could be relocated from the mitochondria to the more protected nucleus. (That is a process evolution has already initiated; only thirteen genes out of the original several thousand remain in the mitochondrion.) Experiments along these lines are being performed at Cambridge.

Another promising path toward radical lifesaving and health-improving measures is *nanomedicine*. Mostly in the planning stages today, nanomedicine will employ tiny mechanical structures that can be configured and programmed to travel to particular sites in the body and perform maintenance and repair tasks. For example, nanotech pioneer Robert A. Freitas, Jr, has designed a nanorobot called a *respirocyte* that would effectively serve as an artificial blood cell. Respirocytes would use sensors to determine levels of particular gases in the body, and pumping mechanisms to distribute oxygen and carbon dioxide as needed. Not only would these artificial cells give heart-attack victims a greater chance at survival (by continuing to oxygenate tissues even when the heart itself has stopped beating), but would also allow people to perform athletic feats that seem superhuman today.[5] Cancer detection is another critical application of nanomedicine and treatment; in 2005, Caltech embarked upon a large-scale research program to study and test nanotech-based chips that detect biomarkers for the presence of various cancers, allowing doctors to monitor the health of specific organs with more precision.[6] In the not too distant future, doctors might routinely use nanomachines not only to detect cancers before they can cause problems, but also to destroy these out-of-control cells without harming surrounding tissues (and without causing massive stress to the body, as contemporary chemotherapy does). Overall, nanomedicine is likely to be an important component of the emerging more effective, less invasive medical paradigm that will make today's methods look comparatively barbaric.

While targeted engineering approaches draw much interest, traditional biogerontology remains an important area of study. After all, the more specific information we gather about anatomy, physiology, and genetics, the more opportunities will present themselves, allowing researchers to study and test new methods of intervention. Laboratory studies on various animals—fruit flies, worms, mice—reveal that life span is not necessarily fixed, but malleable, subject to breeding, direct genetic manipulation, and even diet. One intriguing idea is that if "longevity genes" can be identified in one species, then it might be possible to copy and transfer these genes to another species. Interspecies gene transfer is already quite common in nature; bacteria, for instance, can transfer drug-resistance genes to one another.[7] Clearly, there is much potential for longevity treatments to emerge from a variety of different research avenues, and for these different approaches to learn from one another. Such treatments will surely be in hand by the end of this millennium, if not the end of the century (given the will and the financing), and certainly by the Year Million.

Objections to Superlongevity

Still, however wonderful the prospect of longer, healthier lives might be, a number of issues will invariably be raised while adjusting to this new paradigm. Consider one potential nightmare: the horror of any death in an extremely long-living future civilization, and the paralyzing dread that might stifle initiative and boldness in the "immortalized." Fatal accidents are hard enough for us typical mortals to deal with psychologically. We tend to look for blameworthy causes rather than acknowledge that we live in an indeterminate universe where some accidents can't be avoided; counterintuitively, an explanation after the fact for a particular accident sometimes produces even more anger, guilt, or grief in the survivors. The deaths of young people, young children in particular, are especially difficult for survivors, because the lives of the deceased have been abruptly and unnaturally cut short. Once hundreds or thousands of years of potential thriving life are at stake, death will become even more tragic, meaningless, and perhaps terrifying. One can imagine extreme cases of survivor guilt and an increasing number of people who feel as though if you "live long enough, . . . pretty soon nobody knows you," as one long-lived person remarked to one of us. In essence, you would eventually be known only as the old person you've become (however youthful your body). There wouldn't be anyone left who knew you when you were young, or even during a substantial portion of your life.

Probably most people, given the choice, would still risk survivor guilt or dread for a chance at a longer and healthier life. After all, worrying about being marooned among people who know you only as someone in the last stages of life only makes sense in a culture that is relatively stable and unchanging. In the messy, postindustrial early twenty-first century, people already expect a certain amount of discontinuity, and some thrive on it. Even those who imagine golden ages in the past, or who think the old ways were best, show a remarkable ability to adapt to such newfangled developments as automobiles, airplanes, antibiotics, contact lenses, pacemakers, insulin pumps, PDAs, and cellphones. As the physician and writer Raymond Tallis puts it:

> Those . . . who dislike biotechnology do not seem to realise that the forms of "post-humanity" served up by the natural processes going on in our bodies are a thousand times more radical, more terrifying, and more dehumanizing than anything arising out of our attempts to enhance human beings and their lives. Self-transformation is the essence of humanity,

> and our humanity is defined by our ever-widening distance from the material and organic world of which we are a part, and from which we are apart.[8]

Precisely because many of us lack a clear sense of what the past was really like, we might actually be better prepared for a discontinuous future. If we are heading into an era of biotechnology likely to exceed any previous developments, then what can we expect? And what philosophical difficulties might we encounter?

In the short term, the most pressing moral difficulty in increasing human life span will be making sure that everyone has access to such developments, to provide such benefits on a wide scale, not only to the favored and wealthy few. In a world where average life spans already vary widely among countries, and medical resources are nowhere near being allocated fairly even in developed nations, fairness and equal treatment are unlikely to be the rule, at least at first. Few would argue, though, that nobody should get life saving medical care just because such treatments aren't available to all. It is entirely possible that fear and self-interest will accomplish what altruism might not. Do you really want to be a healthy, well-off, long-lived individual, looking forward to decades of life, in a world where people who don't have your advantages do still have access to plenty of tools that could wreak serious damage on your enclave?

Let's assume our species manages to avoid the worst and embarks on an era of steadily increasing healthy life spans. What problems are most likely to present themselves as we move toward the Year Million? If we learn how to live forever, aside from unpreventable fatal accidents or the choice to terminate one's own existence, would it be wrong to do so?

It's easy to dispense with theological objections to extreme longevity spent in good health because, presumably, *how* a life is lived, and not how *long* it is, decides one's moral status. To be consistent, those who think extending human life necessarily imperils our souls would have to give up on any medical therapy altogether. Very few have yet made that choice, except at the painfully suffering end of life, when only more pain lies ahead—and then, ironically, it is traditional moralists who assert that such people must not be permitted to die by their own choice.

An objection often raised to superlongevity is that long-lived people would get bored. This is easily one of the weakest arguments against learning how to extend life. Some people get bored no matter what diversions are present, while others always find something to interest them. A person's tendency toward (or away from) ennui probably depends on both personality and environmental factors. For example, one "environmental" cause of a tendency toward boredom might be that people accustomed to tight schedules, in which they are constantly running from one obligation to another, never learn how to be still long enough to concentrate on a particular activity and reap the satisfaction it can offer. Perhaps knowing that one has *more* time would alleviate the bad, ennui-promoting mental habits born of experiencing too many demands on too little time.

In any case, if you are going to put forth the boredom argument against superlongevity, you might just as well argue that people should live only to the age of seventy, or some other arbitrarily chosen figure, merely because *some* old people inevitably become victims of ennui. Additionally, in a future where people might be able to alter not only their bodies but also their cognitive abilities and tendencies, there might exist treatments to diminish the tendency toward boredom, much as people today use antidepressant medications to treat chronic depression.

An Overpopulated World?

Must an already overpopulated world become even more crowded, choking an overstressed planet? This assumes that people will continue to have children at current rates. Given that birth rates in more prosperous societies have been dropping already for several generations, there's no reason to assume that they won't continue to drop among longer-lived people. Like older parents now, the healthy and fertile long-lived might be less inclined to have large families, or even to have more than one child. With extended youthfulness, such people are likely to postpone childbearing and rearing, as increasing numbers do now in the developed world.

The planet still might get too crowded, but a more pertinent concern might be the place of children and young people in such a world. Older people might be far more resistant to change and innovation than younger ones, which could create social strains and intergenerational conflicts that will make today's "generation gap" warfare seem like a period of truce or an armistice. Parents who are far older than their children will have different relationships with them than with those closer to them in age, and far fewer couples, straight or gay, are likely to stay together for the entirety of a thousand-year lifetime. But in such a crowded world, becoming parents will be a matter of profound choice and commitment. With the prospect of millennia of healthy life stretching ahead, the twenty years devoted to raising a child will be no great burden. It will more likely be a cause of regret that such a rewarding pursuit is restricted to a brief interval in one's life.

Besides, medically aided longevity might not automatically bring fertility with it. A woman is born with a limited number of ova, which diminish in number from infancy onward; when these are exhausted at menopause, they are gone for good. It's true that advanced medicine might replenish them, but that will be a choice, not an option provided by or an urge imposed by nature. In any case, while we will certainly need creative solutions to deal with overpopulation, surely there has to be a better answer than just condemning to death or nonexistence generations who might otherwise live.

One favorite science-fiction proposal for solving overpopulation is to settle other planets. This solution seems implausible for several reasons. First, and most important, it is highly unlikely that we'd be able to move sufficient people off Earth quickly enough, cheaply enough, and successfully enough to make a dent in the teeming masses that supposedly prompted the need to move in the first place. Second, this proposal assumes there are habitable planets within striking distance, and there simply aren't, not in the solar system, not without a tremendous amount of expensive and time-consuming terraforming. One might also argue that any species careless enough to mess up its original habitat shouldn't get into the business of overrunning other worlds (although such abstractions are probably unlikely to prevent that from happening if it is possible). What of off-Earth orbital habitats, or other such facilities less Promethean than Dyson swarms and Matrioshka Brains? Even with advanced nanotechnology, we might not be able to construct them fast enough to relieve the pressure of natural increase. Besides, these structures might quickly invite their own population booms, as frontiers often do.

Even so, interstellar exploration might be far more important to longer-lived people (and far more interesting), simply because of the immense spacetime distances involved. Many popular entertainments, the *Star Trek* series and the *Star Wars* movies among them, gloss over the vast distances involved in space travel. With known technologies, just moving around our own solar system requires journeys measured in years. Long-lived people would have the time to devote themselves to interstellar space travel and exploration, with a chance to survive until the end of an extended space journey. It's also possible that some travelers would make the trip not in human form (or whatever would pass for human form millennia from now) but as integral parts of space probes and interstellar ships [as discussed by McCarthy, Bradbury, and Hanson in this volume—Ed.]. The notion of uploading one's memories, or having them replicated somehow and integrated into artificial intelligence systems, might have much more appeal after one has been around for a few hundred (or thousand) years and finds oneself looking for some novel activity to pursue. To spread out across the galaxy seems entirely within the means of any species that could live long enough, or suspend life long enough, to reach the goals of their interstellar journeys. Presumably they would also be able to make any biological modifications needed to live on other worlds. Or they might choose to be tourists instead, stopping at various ports of call, although they might more closely resemble itinerant research scientists or restless artists than people on vacation.

Extended Life Span and Species Maturity

Arguably, a pervasive lack of long-term thinking is a major contributing factor to our species' current perilous situation. Enabling a greater capacity for long-term thinking might prove the most valuable gift extended life can give us. One intriguing piece of psychological research, cited by Robert A. Freitas, Jr., tested human subjects' reactions to expanded or contracted time frames. "Subjects with no future experienced a loss of identity and a profound euphoric mystical sensation. . . . Expanded futures canceled all fear of death, inducing serene calmness and happiness."[9] Intuitively, this makes sense. Lack of time to devote to important tasks, family or friends, and recreation is among the most common complaints of people in countless surveys.

Many in the West have already radically changed their attitudes toward old age. Just how radically might not be obvious to people too young to remember a time when it would have been considered eccentric at best for sixty- and seventy-year-olds to be out jogging, attending and performing at rock concerts, and wandering around in the kinds of clothing once worn only by teenagers. People fortunate and wealthy enough are already rethinking what to do with their lives in their seventies and eighties; they no longer automatically assume that they will be either dead or decrepit by that point. Whatever problems longer life might bring us, lack of time and carelessness about the long-term effects of our actions are not likely to be among them.

Paradoxically, as our species matures, we might have time to become childlike in the best sense of that word—open, curious, and receptive to whatever new delights the universe has to offer. There would be enough time for us to create and cherish those new breeds of flowers, pets, and other life-forms so charmingly evoked by Freeman Dyson.[10] We would no longer need to think of "retirement" but rather of periodic, years-long vacations between entire careers or long-term projects. We could learn all the world's (or perhaps worlds') languages, and perhaps even invent our own. We could write poems and compose songs of such sweeping scope and subtlety that they might take hundreds or thousands of years to write. We could spend aeons exploring the depths of mathematics, and then a few more studying biology, and then still more developing entertainment. And because the universe itself will continue to evolve and change even as we change, the very notion of boredom will become antiquated.

There would at last be time enough, in short, for us to discover more completely who and what we are, to acquire a personal history that also encompasses a larger history, to look forward to a future that will give us second, third, fourth—and more—chances to get it right, finally . . . or at least to patch up our mistakes. No, we wouldn't mind living until the Year Million, or forever. Especially given the alternative.

Chapter 10

Communicating with the Universe

Amara D. Angelica

Over the next million years, a descendant of the Internet will maintain contact with inhabited planets throughout our galaxy and begin to spread out into the larger universe, linking up countless new or existing civilizations into the Universenet, a network of ultimate intelligence.

The Earth has already input information into the Universenet. Whenever microwave towers or satellites send Internet traffic, some of the energy leaks off, transmitting data unintentionally into space. The first e-mail messages transmitted via microwave towers in 1969 by the predecessor of the Internet, ARPANET, have (theoretically) traveled thirty-nine light-years so far, way past the nearest star system, Alpha Centauri, four light-years away. In practice, such feeble signals are probably buried in cosmic radio noise.

Now NASA plans to do it intentionally. The Interplanetary Internet (IPI) should allow NASA to link up the Internets of Earth, spacecraft, and eventually the Moon, Mars, and beyond.[1] By the Year Million, billions of "smart dust" sensors will be connected to a distant descendant of the IPI, exchanging data in real time or via store-and-forward protocol or wireless mesh (a network that handles many-to-many connections and is capable of dynamically updating and optimizing these connections) on planets and in spacecraft.[2]

Meanwhile, one important near-future use will be for tracking asteroids, comets, and space junk, exchanging three-dimensional position location and time data (similar to GPS on Earth) via multiple hops between sensors. Once affordable personal space travel is available, the IPI could serve as the core of an interplanetary version of air traffic control. The IPI scheme could also become the standard communications protocol as we expand out beyond the solar system's planets, and then beyond the stars and to other galaxies. We could start with possibly habitable planets beyond the solar system, such as Gliese 581d, the third planet of the red dwarf star Gliese 581 (about twenty light-years away from Earth), if we detect signs of intelligent life there.

But using radio waves or lasers to communicate with civilizations around other stars, let alone in other galaxies, requires huge amounts of energy. Exactly how much energy? That depends mainly on distance, frequency, directional efficiency of antennas, and assumed ability of the receiving civilization to detect signals amid the extreme electromagnetic noise of space. In 1974, the Arecibo telescope beamed a 210-byte radio message aimed at the globular star cluster M13, some twenty-five thousand light-years away. It was transmitted with a power of one megawatt—enough energy to power about one thousand homes, using a narrow beam to achieve an EIRP (effective isotropic radiated power) of twenty trillion watts. That made it the strongest human-made signal ever sent. (It has gone 0.14 percent of the way, so far.)

Arecibo uses a large dish. Another way to create a narrow beam of high-power microwave radio energy is to build a phased-array antenna with multiple dishes spread out over a large area. These could be located on the Moon or at a Lagrange point (one of the stable locations in the Earth-Moon-Sun axis). Or a high-powered laser could be used. How highly powered? Looking toward the Year Million, as we reach out to communication nodes orbiting more distant stars, or in other galaxies, we will need to use a *lot* of power—as much as the entire power of the Sun. A civilization able to do that kind of cosmic engineering is referred to as Kardashev Type II, or KT-II [see Chapter 8—Ed.].

By modest contrast, our civilization used about fifteen terawatt-hours in 2004 (a terawatt-hour is one billion kilowatt-hours) of electrical power.[3] New York University physics professor emeritus Martin Hoffert and other scientists calculate that if our power consumption grows by just 2 percent per year, then in just four hundred years we will need all the solar power received by the Earth (10^{16} watts = 10,000 terawatts). And in a thousand years, we'll require all of the power of the Sun (4×10^{26} watts).[4] Hoffert and other scientists propose space-based solar power as one major future solution. Solar flux is eight times higher in space than the surface average on cloudy Earth and available twenty-four hours a day, unlike solar energy panels on Earth. Power satellites located in geosynchronous orbit (like communication satellites) would use a bank of photovoltaic receptors to convert the Sun's energy to radio waves. This energy would be beamed wirelessly down to a large "rectenna," or rectifying antenna, where the incoming microwave energy is rectified (converted) for use in the electrical power grid on Earth, turning it into electricity for distribution. Alternatively, laser beams could replace radio-frequency signals.[5] Once the infrastructure is in place for economically launching space-based solar power satellites, the same types of microwave or laser systems could be aimed at the stars for communicating elsewhere.

Eventually, when we have become first a KT-I and then a KT-II civilization, we will reach even farther out, to supergalaxies and even to clusters of supergalaxies, which could require a Type III civilization—one capable of controlling the power of an entire galaxy, some 10^{36} watts. The communication latencies (transmission delays) for such a system would be millions, or even hundred of millions, of years. (Two-way latency is already a problem for astronauts in the solar system, increasing as we transmit information to places farther from the Earth, or wherever humans and posthumans end up, perhaps uploaded into a Matrioshka Brain that will have replaced the existing solar system.) Even the nearest star, Alpha Centauri, could not reply to a message sooner than eight years after it was sent. Talk about bad netiquette.

Possibly the denizens of the Year Million will solve this time lag with extreme cosmic engineering feats such as wormholes, or even communication via parallel universes.[6] One intriguing possibility is the use of quantum entanglement—that is, allowing an entangled atom or photon to carry information across a distance, theoretically anywhere in the universe (once the initial photons have been received), or "spooky action-at-a-distance," as Einstein called it.[7] An experiment testing the possibility of communication using this principle is in progress in the Laser Physics Facility at the University of Washington by Professor John G. Cramer.[8] Cramer astonished physicists at a joint American Institute of Physics/American Association for the Advancement of Science conference in 2006 by presenting experimental evidence that the outcome of a laser experiment could be affected by a future measurement: a message was sent to a time fifty microseconds in the past.[9] This leads to an even more bizarre idea: retrocausal communication—the future affecting the past, as theoretical physicist Jack Sarfatti (the inspiration for Doc in the movie *Back to the Future*) has proposed.[10] So in principle, perhaps one could bypass the speed-of-light limitation and have messages show up in a distant galaxy long before they could have been received by radio or laser transmission, or even before they were sent!

Web to ET: Download This

Humans might not be the first technological species to explore the galaxy. Suppose alien probes await us in orbit or on the Moon (like the obelisk in Arthur C. Clarke's *The Sentinel* and its movie version, *2001: A Space Odyssey*) or at Lagrange points.[11] If so, we might only need to respond with the right signals to trigger a connection—similar to logging on to an FTP server with the right IP address, user name, and password. Such probes might even now be scattered around the solar system as smart dust particles that we haven't yet analyzed. IBM has developed a prototype of a molecular switch that could replace current silicon-based chip technology with atom-based processors, making it theoretically possible to run a supercomputer on a chip the size of a speck of dust. IBM is also developing technology to store a bit on a single atom, portending hard drives that can pack up to a thousand times as much information on a hard disk as current technologies.[12]

Instead of transmitting via radio or laser, sending a physical data spore might be a simpler and more effective alternative. Rutgers University electrical engineer Christopher Rose has shown that for long messages conveyed across long distances (where transmitting a signal would be extremely expensive, have limited range, or be too hard to find), it is more effective to send physical messages than transmit them. That was one rationale for sending the greeting plaque on Pioneer 10 and 11 in 1972 and 1973, and a more complex inscribed disk on the Voyager probe in 1977. Rose thinks there could be such inscribed objects now orbiting planets in our solar system, or on asteroids.[13]

But transmitting information into space still fires up the imagination of several scientists. SETI senior astronomer Seth Shostak has proposed that rather than sending simple coded messages, why not just feed the Google servers into the transmitter and send the aliens the entire Web? It would take about half a year to transmit the Web in the microwave region at one megabyte a second; infrared lasers operating at a gigabyte per second would shorten the broadcast time to no more than two days.[14] Transmitting the Web into space could also serve as a backup for civilization. William E. Burrows has suggested creating a self-sufficient colony on the Moon where a "backup drive" could store the history and wisdom of civilization in case a calamity strikes Earth.[15] To achieve this, Burrows set up an organization, Alliance to Rescue Civilization (ARC), subsequently absorbed by the Lifeboat Foundation, which is developing solutions to prevent the extinction of mankind. Acquiring knowledge from ancient extraterrestrial civilizations could be critical to our long-term human survival, says Lifeboat Foundation president Eric Klien. "The Universenet could give us the final signals of a civilization right before it destroyed itself," he wrote in a Skype message. "We could use this information to avoid our own destruction, perhaps the most important reason to continue the SETI project. If we learned that a civilization was destroyed by, say, nanoweapons, we could start creating defenses against this situation."[16]

Such signals might not be obvious. For example, pulsars, discovered in 1967, are rotating neutron stars that emit electromagnetic waves. Their rapid rotation causes their radiation to become pulsed. Could this radiation be modulated deliberately to form a sort of cosmic transmitter? Astronomers at first thought the pulses meant they might be ET; so far, they haven't found any evidence of an actual message. Or are there encoded messages too subtle to detect? And pulsars are far from the most powerful possible signal sources from space. Quasars can release the energy equal to hundreds of average galaxies combined, equivalent to one trillion Suns. Could they be galactic Web sites run by Type III civilizations? (Unlikely, since most quasars are very far away, which means distant in time, and seem to have been formed not long after the emergence of the universe from the Big Bang.)

Computer scientist Stephen Wolfram believes current methods used in SETI are inefficient and unlikely to produce reliable results because our detection methods seek to detect only *regular* patterns. A more efficient method would use sophisticated, noise-immune coding, producing something similar to spread spectrum signals. To SETI's present system of analysis this kind of signal sounds and looks like random noise, and would be overlooked and discarded.[17] Wolfram suggests we need more sophisticated software-based signal processing. Maybe we need someone like Hedy Lamarr, the brilliant actress who famously said, "Any girl can be glamorous; all she has to do is stand still and look stupid," and then went on to invent spread spectrum technology. Could ET be using it? There's no way to know with the current SETI technology. Complex artifacts made by an advanced civilization could look very much like natural objects, Wolfram argues. Could the stars themselves be extraterrestrial artifacts? "They could have been built for a purpose," says Wolfram. "It's extremely difficult to rule it out."[18]

Is Alien Intelligence Hidden in Junk DNA?

Cardiff University astronomer and mathematics professor Chandra Wickramasinghe, a longtime collaborator with the late cosmologist Sir Fred Hoyle, has suggested that life on this planet began on comets, since their combination of clay and water is an ideal breeding ground for life. He believes that explanation to be a quadrillion times more likely than Earth's having spawned life.[19] If that's the case, then comets and asteroids could be carrying physical messages, a sort of "sneakernet"—physical file sharing in the interests of added security—for the Universenet.

Astrobiologist Paul Davies, now at Arizona State University, suggests that ET could embed messages in highly conserved sections of viral DNA—most likely in its so-called "junk" sections—and send them out as hitchhikers on asteroids or comets. (Genomics researchers at the Lawrence Berkeley National Laboratory in California, who compared human and mouse DNA, have reported millions of base pairs of highly conserved sequences of junk DNA, meaning they have a survival value.)[20] These messages could even have been incorporated into terrestrial life, Davies thinks, and lurk in our DNA, awaiting interpretation. (There could be an interesting d-mail—DNA-mail—waiting to be discovered as we search through the decoded genome.) Rather than beaming information randomly in the hope that somewhere, someday, an intelligent species will decode them, this method would use a pre-existing "legion of small, cheap, self-repairing and self-replicating machines that can keep editing and copying information and perpetuate themselves over immense durations in the face of unforeseen environmental hazards. Fortunately, such machines already exist. They are called living cells."[21]

Transmitting People to the Stars

Futurist/inventor Ray Kurzweil has suggested that once the intelligent life on a planet invents machine computation, it is only a matter of a few centuries before its intelligence saturates the matter and energy in its vicinity. At that point, he suggests, nanobots will be dispersed like the spores of plants. This colonization will eventually expand outward, approaching the speed of light [as discussed in Chapter 9—Ed.].[22] In Fred Hoyle and John Elliot's 1962 novel *A For Andromeda,* a radio signal from the direction of the galaxy M31 in Andromeda gives scientists a computer program for the creation of a living organism, adapting borrowed human DNA. They name this young cloned woman Andromeda, and through her agency the computer tries to take over the world.[23] Author James Gardner has seriously suggested a version of such "interstellar cloning": an advanced civilization could transmit a software program to us with instructions on replicating its own inhabitants—even an entire civilization.[24]

Dr. Martine Rothblatt, who founded Sirius Satellite and other satellite companies, has suggested a related method for connecting with Universenet: sending *bemes* or *units of being*—highly individual elements of personality, mannerisms, feelings, recollections, beliefs, values, and attitudes. Bemes are fundamental, transmittable, mutable units of beingness in the spirit of *memes* (Richard Dawkins's term for the replicators of cultural information that a mind transmits, verbally or by demonstration, to another mind). The main difference is that memes are culturally transmittable elements that have *common* meanings, whereas bemes reflect *individual* characteristics.

Rothblatt suggests that a new Beme Neural Architecture (BNA) will outcompete DNA in populating the universe. "At any moment, and certainly at some moment, a giant star in our general stellar neighborhood will blow up and thereby fry everything in its vicinity," she points out. Some of these explosions, known as gamma-ray bursts, are so violent that they damage everything within hundreds of light-years. Yet there are two or three gamma-ray bursts somewhere in the observable universe *every day* and about one thousand less explosive but still life-ending supernovas every day throughout the galaxies (that we can observe). One explanation for the Fermi Paradox—why there is no evidence of ET, although the galaxy seems capable of so many extraterrestrial civilizations—is that sooner or later a supernova nabs everyone's life zone. "Perhaps the only way we can survive the risk of astrobiological or mega-volcanic catastrophe is to spread ourselves out among the stars," Rothblatt suggests. And as self-replicating code, bemes are much more quickly assembled, replicated, and transported than genes strung along chromosomes and transmitted by sex. Computer technology is vastly more efficient than wet biology in copying information. Expressed in digital bits rather than in nucleotide base pairs, information can be transported farther (beyond Earth to evade killer asteroid impacts) and faster (at the speed of light).

DNA is not well suited for space travel. It can replicate effectively only within bodies. Humans require vast quantities of life-preserving supplies, and besides, at the moment, we don't live long enough to make the journey to other stars [a factor, as Pamela Sargent notes in the previous chapter, that is subject to change—Ed.]. On the other hand, by replicating our minds into BNA and storing them in a computer substrate, we can travel far longer and far faster, since we would be traveling with minimal mass in seeds or spores. Arriving at a promising planet, our BNA can be loaded into nanotech-built machine bodies to prepare a new home. Once that home flourishes, human (and other) DNA can be reconstructed from either stored samples or digital codes and basic chemicals, which can be nurtured into mature bodies free to develop their own minds or to receive a transfer of a BNA mind.

Alternatively, Rothblatt suggests that just by spacecasting your bemes, you can already achieve a level of immortality, and so can all of humanity. In March 2007, Rothblatt's CyBeRev Project began experimentally spacecasting bemes in the form of digitized video, audio, text, personality tests, and other recordings, of attributes of a person's being, such as memories, mannerisms, personality, feelings, recollections, beliefs, attitudes, and values.[25] These bemes are transmitted out into the universe via a microwave dish normally used to communicate with satellites. Any spacecast signal, she speculates, has a chance of being decoded from the background cosmic noise in the same way a cellphone's CDMA (spread spectrum) encoded signal is decoded out of random electromagnetic noise. Your bemes could then be interpreted, and you yourself re-created from the transmission. This requires interception by an advanced, intelligent civilization that would receive and decode the signals, then instantiate the bemes as either a regenerated traditional cellular or a bionanotechnological, body. (If this happened, we might find by the Year Million that the galaxy is swarming with other humans downloaded by far-flung extraterrestrials or their machines.)

Each spacecast of an individual's bemes is accompanied by an informed-consent form authorizing that individual's re-instantiation from the transmitted bemes. The CyBeRev project is based on the hypothesis that advanced intelligence will respect sentient autonomy and be capable of filling in the blanks of a person's consciousness via interpolation of the spacecast bemes, using background cultural information transmitted from Earth. The project's backers do not believe extraterrestrials will unethically revive persons such as television personalities, whose images, behavior, and personal information have been telecast, but who have not authorized their re-instantiation. Still, such cultural transmissions will be useful in the aggregate, providing revived spacecasters with a familiar environment. "Given the vast amount of television and Internet information streaming into space, the revivers of our spacecasters will have abundant contextual information with which to work," concluded Rothblatt.[26]

Programming the Universe

By converting matter into what some futurists call *computronium* (hypothetical material designed to be an optimized computational substrate), Year Million scientists could create the beginnings of an ultimately powerful computer.[27] Taking it to the extreme, MIT scientist Seth Lloyd has calculated that a computer made up of all the energy in the entire known universe (that is, within the visible "horizon" of forty-two billion light-years) can store about 10^{92} bits of information and can perform 10^{105} computations/second.[28] The universe itself is a quantum computer, he says, and it has made a mind-boggling 10^{122} computations since the Big Bang (for that part of the universe within the "horizon").[29] Compare that to about 2×10^{28} operations performed over the entire history of computation on Earth ("because of Moore's law, half of this computation has taken place in the last year and a half," he wrote in 2006). What's more, the observable horizon of the universe, space itself, is expanding at three times the speed of light (in three dimensions), so the amount of computation performable within the horizon increases over time.

Lloyd has also proposed that a black hole could serve as a quantum computer and data storage bank. In black holes, he says, Hawking radiation, which escapes the black hole, unintentionally carries information about material inside the black hole. This is because the matter falling into the black hole becomes entangled with the radiation leaving its vicinity, and this radiation captures information on nearly all the matter that falls into the black hole. "We might be able to figure out a way to essentially program the black hole by putting in the right collection of matter," he suggests.[30]

There is a supermassive black hole in the center of our galaxy, perhaps the remnant of an ancient quasar. Will this become the mainframe and central file-sharing system for galaxy hackers of the Year Million? What's more, a swarm of ten thousand or more smaller black holes may be orbiting it.[31] Might they be able to act as distributed computing nodes and a storage network? Toward the Year Million, an archival network between stars and between galaxies could develop an *Encyclopedia Universica*, storing critical information about the universe at multiple redundant locations in those and many other black holes.

Clash of the Titans

Far beyond the Year Million, our galaxy faces a crisis. The supermassive black holes in our galaxy and the Andromeda galaxy are headed for a cosmic collision in two billion years. Will they have incompatible operating systems—a sort of

Mac-versus-PC confrontation? (Of course, they might just pass by each other—or be steered past by hyperintelligent operators.)

In *The Intelligent Universe*, James Gardner adapted a bold notion originally proposed by cosmologist Lee Smolin. For Smolin, Darwinian principles constrain the nature of any universe such that new baby universes produced via black holes will resemble their parent cosmos, and will be surprisingly life-friendly as well. Gardner extends this idea into a fundamentally radical (but falsifiable) hypothesis called the Selfish Biocosm—the cosmological equivalent of Richard Dawkins's selfish gene. The idea is that eventually intelligent life must acquire the capacity to shape the entire cosmos. In addition, the universe has a Smolin-style "utility function": propagation of baby universes exhibiting the same life-friendly physical qualities as their parent universe, including a system of physical laws and constants that enables life and intelligence to emerge and eventually repeat the cycle.

Under this scenario, the mission of sufficiently evolved intelligent life in the universe is to serve as a cosmic reproductive organ—the equivalent of DNA in living creatures—spawning an endless succession of life-friendly offspring that are themselves endowed with the same reproductive capacities as their predecessors. (Rothblatt's BNA might well be the fundamental mechanism for this evolutionary process—veteran physicist John Wheeler's legendary IT from BIT, things arising from information rather than the other way round.)

Gardner believes that we've already received a message from ET: the laws and constants of our universe, including the inexplicable cosmological constant which at this time is accelerating cosmic expansion. His hypothesis makes sense of the observation that the constants seem rigged in favor of the emergence of life. For example, they are improbably hospitable to carbon-based intelligent life—an unlikely and as yet unexplained anthropic oddity that some scientists have identified as the deepest mystery in all of science. As Gardner claims:

> We are likely not alone in the universe, but are probably part of a vast—yet undiscovered—transterrestrial community of lives and intelligences spread across billions of galaxies and countless parsecs. . . . We share a possible common fate with that hypothesized community: to help shape the future of the universe and transform it from a collection of lifeless atoms into a vast, transcendent mind.

In the Year Million, such a cosmic community will be linked up by the Universenet.

Chapter 11

The Great Awakening

Rudy Rucker

Ubiquitous Nanomachines

Molecular nanotechnology is the craft of manufacturing things on the molecular scale. One goal is to create programmable nanobots: tailor-made agents roughly the size of biological viruses. The comparison is apt. What's likely to play out is that, over the coming centuries and millennia, we'll be capitalizing on the fact that biology is already doing molecular fabrication. The nascent field of synthetic biology is going to be the true nanotech of the future.

One immediate worry is what nanotechnologists have called the "gray goo problem." That is, what's to stop a particularly virulent, artificial organism from eating everything on Earth? My guess is that this could never happen. Every existing plant, animal, fungus, and protozoan *already* aspires to world domination. There's nothing more ruthless than viruses and bacteria—the grizzled homies who've thrived by keeping it real for some three billion years.

The fact that artificial organisms are likely to have simplified metabolisms doesn't necessarily mean that they're going to be faster and better. It's more likely that they'll be dumber and less adaptable. My sense is that, in the long run, Mother

Nature always wins. Cautionary note: Mother Nature's "win" may not include the survival of the pesky human race!

But let's suppose that all goes well and we learn to create docile, biological nanobots. There's one particular breed that I like thinking about; I call them *orphids.*

The way I imagine it, orphids reproduce using ambient dust for raw material. They'll cover Earth's surface, yes, but they'll be well behaved enough to stop at a density of one or two orphids per square millimeter, so that you'll find a few million of them on your skin and perhaps ten sextillion orphids on Earth's whole surface. From then on, the orphids reproduce only enough to maintain that same density. You might say they have a conscience, a desire to protect the environment. And, as a side benefit, they'll hunt down and eradicate any evil nanomachines that anyone else tries to unleash.

Orphids use quantum computing; they propel themselves with electrostatic fields; they understand natural human language. One can converse with them quite well. I'll suppose that an individual orphid is roughly as smart as a talking dog with, let us say, a quadrillion bytes of memory being processed at a quadrillion operations per second.

How do we squeeze so much computation out of a nanomachine? Well, a nanogram does hold about a trillion particles, which gets us close to a quadrillion. According to quantum physicist Seth Lloyd, if we regard brute matter as a quantum computation, then we do have some ten quadrillion bytes per nanogram.[1] So there's only, *ahem*, a few implementation details in designing molecular nanomachines smart enough to converse with.

The orphids might be linked via electromagnetic wireless signals that are passed from one to the next; alternatively, they might use, let us say, some kind of subdimensional faster-than-light quantum entanglement. In either case, we call the resulting network the Orphidnet.

Omnividence and Telepathy

We can suppose that the orphids will settle onto our scalps like smart lice. They'll send magnetic vortices into our occipital lobes, creating a wireless human interface to the Orphidnet. Of course, we humans can turn our connection on and off, and we'll have read-write control. As the Orphidnet emerges, we'll get intelligence amplification.

So now everyone is plugged into the Orphidnet all the time. Thanks to the orphid lice, everyone has a heads-up display projected over the visual field. And thanks to global positioning systems, the orphids act as tiny survey markers—or as the vertices of computer-graphical meshes. Using these real-time meshes, you actually see the shapes of distant objects. The orphids will be sensitive to vibrations, so you can hear as well. We'll have complete *omnividence*, as surely as if the earth were blanketed with video cameras.

One immediate win is that violent crime becomes impossible to get away with. The Orphidnet remembers the past, so anything can be replayed. If you do something bad, people can find you and punish you. Of course someone *can* still behave like a criminal if he holds incontrovertible physical force—if, for instance, he is part of an armed government. I dream that the Orphidnet-empowered public sees no further need for centralized and weaponized governments, and humankind's long domination by ruling elites comes to an end. Another win is that we can quickly find missing objects.

The flip side of omnividence is that nobody has any privacy at all. We'll have less shame about sex; the subject will be less shrouded in mystery. But sexual peeping will become an issue, and as omnividence shades into telepathy, some will want to merge with lovers' minds. But surely lovers can find some way to shield themselves from prying. If they can't actually turn off their orphids, the lovers may have physical shields of an electromagnetic or quantum-mechanical nature; alternately, people may develop mantralike mental routines to divert unwanted visitors.

Telepathy lies only a step beyond omnividence. How will it feel? One key difference between omnividence and telepathy is that telepathy is participatory, not voyeuristic. That is, you're not just watching someone else; you're picking up the person's shades of feeling.

One of the key novelties attending electronic telepathy is the availability of psychic hyperlinks. Let me explain: Language is an all-purpose construction kit that a speaker uses to model mental states. In interpreting these language constructs, a listener builds a mental state similar to the speaker's. Visual art is another style of construction kit; here an idea is rendered in colors, lines, shapes, and figures.

As we refine our techniques of telepathy, we'll reach a point where people converse by exchanging hyperlinks into each other's mind. It's like sending someone an Internet link to a picture on your Web Site—instead of sending a pixel-by-pixel copy of the image. Rather than describing my weekend in words, or showing you pictures that I took, I simply pass you a direct link to the memories in my head. In other words, with telepathy, I can let you directly experience my thoughts without my explaining them via words and pictures. Nevertheless, language will persist. Language is so deeply congenial to us that we'd no sooner abandon it than we'd give up sex.

On a practical level, once we have telepathy, what do we do about the sleazeball spammers who'll try to flood our minds with ads, scams, and political propaganda? We'll use adaptive, evolving filters. Effective spam filters behave like biological immune systems, accumulating an ever-growing supply of "antibody" routines. In a living organism's immune system, the individual cells share the antibody techniques they discover. In a social spam filter, the individual users will share their fixes and alerts.

Another issue with telepathy has to do, once again, with privacy. Here's an analogue: a blogger today is a bit like someone who's broadcasting telepathically, dumping his or her thoughts into the world for all to see. A wise blogger censors his or her blog, so as not to appear like a hothead, a depressive, or a bigot.

What if telepathy can't be filtered, and everyone can see everyone else's secret seething? Perhaps, after a period of adjustment, people would get thicker skins. Certainly it's true that in some subcultures, people yell at each other without necessarily getting excited. Perhaps a new kind of tolerance and empathy might emerge, whereby no one person's internal turmoil seems like a big deal. Consider: to be publicly judgmental of someone else, you compare your well-tended *outside* to the other person's messy *inside.* But if everyone's insides are universally visible, no one can get away with being hypocritical.

Telepathy will provide a huge increase in people's ability

to think. You'll be sharing your memory data with everyone. In the fashion of a Web search engine, information requests will be distributed among the pool of telepaths without the need for conscious intervention. The entire knowledge of the species will be on tap for each individual. Searching the collective mind won't be as fast as retrieving something from your own brain, but you'll have access to far more information.

Even with omnividence and telepathy, I expect that, day in and day out, people won't actually change that much—not even in a million years. That's a lesson history teaches us. Yes, we've utterly changed our tech since the end of the Middle Ages, but the paintings of Hieronymus Bosch or Peter Bruegel show that people back then were much like us, perennially entangled with the seven deadly sins.

No matter the tech, what people do is based upon simple needs: the desire to mate and reproduce, the need for food and shelter, and the longing for power and luxuries. Will molecular manufacture give all of us the luxuries we want? No. Skewed inverse power-law distribution of valued qualities is an intrinsic property of the natural world. That is, roughly speaking, if there are a thousand people at the bottom of the heap, and a hundred immediately above them, there'll be only ten farther up, and just one perched on the top in possession of a large proportion of the goodies. Even if we become glowing clouds of ectoplasm, there's going to be something that we're competing for—and most of us will feel as though we're getting screwed.

Those goodies need not be "possessions" as we understand them; in the near term, an interesting effect will emerge. Since we're all linked on the net, we can easily borrow things, or even get things free. As well as selling things, people can loan them out or give them away. Why? To accumulate social capital and good reputation.

In the Orphidnet future, people can always find leftover food. Some might set out their leftovers, like pies for bums. Couch surfing as a serial guest becomes eminently practical, with the ubiquitous virtual cloud of observers giving a host some sense of security vis-à-vis the guests. And you can find most of the possessions you need within walking distance—perhaps in a neighbor's basement. A community becomes a shared storehouse.

On the entertainment front, I imagine Orphidnet reality soap operas. These would be like real-time video blogs, with sponsors' clickable ads floating around near the characters, who happen to be interesting people doing interesting things.

People will still dine out—indeed this will be a preferred form of entertainment, as physically eating something is one of the few things that require leaving the home. As you wait at your restaurant table for your food, you might enjoy watching (or even experiencing) the actions of the chef. Maybe the restaurant employs a gourmet eater, with such a sensitive and educated palate that it's a pleasure to mind meld when this eater chows down.

Will telepaths get drunk and stoned? Sure! And with dire consequences. Imagine the havoc you could wreak by getting wasted and "running your brain" instead of just emailing, phoning, or yelling at people face-to-face. There will be new forms of intoxication as well. A pair of people might lock themselves into an intense telepathic feedback loop, mirroring their minds back and forth until chaotic amplification takes hold.

In the world of art, suppose someone finds a way to record mood snapshots. Then we can produce objects that directly project the raw experiences of transcendence, wonder, euphoria, mindless pleasure, or sensual beauty, without actually having any content.

Telepaths will use language for superficial small talk, but, as I mentioned, they'll just as often use psychic hyperlinks and directly exchanged images and emotions. Novels could take the form of elaborate sets of mental links. Writing might become more like video blogging. A beautiful state of mind could be saved into a memory network, glyph by glyph. This new literary form might be called the metanovel.

Artificial Intelligence and Intelligence Amplification

In the ubiquitous nanobot model I've been discussing, the Orphidnet, we have a vast array of small linked minds. It's reasonable to suppose that, as well as helping humans do things, the Orphidnet will support emergent, artificially intelligent agents that enlist the memory and processing power of a few thousand or more individual orphids.

Some of these agents will be as intelligent as humans, and some will be even smarter. It's easy to imagine them being willing to help people by carrying out such tasks as complex and tedious searches for information or by simulating and evaluating multiple alternate action scenarios. The result is that humans would undergo IA, or intelligence amplification.

A step further, intelligent Orphidnet agents group into higher minds that group into still higher minds and so on, with one or several planetary-level minds at the top. Here, by the way, is a fresh opportunity for human excess. Telepathically communing with the top mind will offer something like a mystical experience or a drug trip. The top mind will be like a birthday piñata stuffed with beautiful insights woven into ideas that link into unifying concepts that puzzle piece themselves into powerful systems that are in turn aspects of a cosmic metatheory—*aha*! Hooking into the top mind will make any individual feel like more than a genius. Downside: once you unlink you probably won't remember many of the cosmic thoughts that you had, and you're going to be too drained to do much more than lie around for a few days.

Leaving ecstatic merging aside, let's say a little more about intelligence amplification. Suppose that people reach an effective IQ of 1,000 by taking advantage of the Orphidnet memory enhancement and the processing aid provided by the Orphidnet agents. Let's speak of these kilo-IQ people as ***kiqqies***.

As kiqqies, they can browse through all the world's libraries and minds, with Orphidnet agents helping to make sense of it all. How would it feel to be a kiqqie?

I recently had an e-mail exchange about this with my friend Stephen Wolfram, a prominent scientist who happens to be one of the smartest people I know. When I asked him how it might feel to have an IQ of 1,000, and what that might even mean, he suggested that the difference might be like that between simulating something by hand and simulating it on a high-speed computer with excellent software. Quoting from Wolfram's e-mail:

> There's a lot more that one can explore, quickly, so one investigates more, sees more connections, and can look more moves ahead. More things would seem to make sense. One gets to compute more before one loses attention on a particular issue, etc. (Somehow that's what seems to distinguish less intelligent people from more intelligent people right now.)

Against Computronium

In some visions of the far future, amok nanomachines egged on by corporate geeks are disassembling the solar system's planets to build Dyson shells of computronium around the Sun. Computronium is, in writer Charles Stross's words, "matter optimized at the atomic level to support computing." A Dyson shell [as several previous essays have discussed—Ed.] is a hollow sphere of matter that intercepts all of the central Sun's radiation—using some of it and then passing the rest outwards in a cooled-down form, possibly to be further intercepted by outer layers of Dyson shells. What a horrible thing to do to a solar system!

I think computronium is a spurious concept. Matter, just as it is, conducts outlandishly complex chaotic quantum computations by dint of sitting around. Matter isn't dumb. Every particle everywhere and every when computes at the max possible flop. I think we tend to very seriously undervalue quotidian reality.

Turning an inhabited planet into a computronium Dyson shell is comparable to filling in wetlands to make a mall, clear-cutting a rain forest to make a destination golf resort, or killing a whale to whittle its teeth into religious icons of a whale god.

Ultrageek advocates of the computronium Dyson-shell scenario like to claim that nothing need be lost when Earth is pulped into computer chips. Supposedly the resulting computronium can run a VR (virtual reality) simulation that's a perfect match for the old Earth. Call the new one Vearth. It's worth taking a moment to explain the problems with trying to replace real reality with virtual reality. We know that our present-day videogames and digital movies don't fully match the richness of the real world. What's not so well known is that no feasible VR can *ever* match nature, because there are no shortcuts for nature's computations. Due to a property of the natural world that I call the "principle of natural unpredictability," fully simulating a bunch of particles for a certain period of time requires a system using about the same number of particles for about the same length of time. Naturally occurring systems don't allow for drastic shortcuts.[2]

Natural unpredictability means that if you build a computer-simulated world that's smaller than the physical world, the simulation cuts corners and makes compromises, such as using bitmapped wood-grain, linearized fluid dynamics, or cartoon-style repeating backgrounds. Smallish simulated worlds are doomed to be dippy Las Vegas/Disneyland environments populated by simulated people as dull and predictable as characters in bad novels.

But wait—if you *do* smash the whole planet into computronium, then you have potentially as much memory and processing power as the intact planet possessed. It's the same amount of mass, after all. So then we *could* make a fully realistic world-simulating Vearth with no compromises, right? Wrong. Maybe you can get the hardware in place, but there's the vexing issue of software. Something important goes missing when you smash Earth into dust: you lose the information and the software that was embedded in the world's behavior. An Earth amount of matter with no high-level programs running on it is like a potentially human-equivalent robot with no AI software, or, more simply, like a powerful new computer with no programs on the hard drive.

Ah, but what if the nanomachines first copy all the patterns and behaviors embedded in Earth's biosphere and geology? What if they copy the forms and processes in every blade of grass, in every bacterium, in every pebble—like Citizen Kane bringing home a European castle that's been dismantled into portable blocks, or like a foreign tourist taking digital photos of the components of a disassembled California cheeseburger?

But, come on, if you want to smoothly transmogrify a blade of grass into some nanomachines simulating a blade of grass, then why bother grinding up the blade of grass at all? After all, any object at all can be viewed as a quantum computation! The blade of grass already *is* an assemblage of nanomachines emulating a blade of grass. Nature embodies superhuman intelligence just as she is.

Why am I harping on this? It's my way of leading up to one of the really wonderful events that I think our future holds: the withering away of digital machines and the coming of truly ubiquitous computation. I call it the Great Awakening.

I predict that eventually we'll be able to tune in telepathically to nature's computations. We'll be able to commune with the souls of stones.

The Great Awakening will eliminate nanomachines and digital computers in favor of naturally computing objects. We can suppose that our newly intelligent world will, in fact, take it upon itself to crunch up the digital machines, frugally preserving or porting all of the digital data.

Instead of turning nature into chips, we'll turn chips into nature.

The Advent of Panpsychism

In the future, we'll see all objects as alive and conscious—a familiar notion in the history of philosophy and by no means disreputable. Hylozoism (from the Greek *hyle*, matter, and *zoe*, life) is the doctrine that all matter is intrinsically alive, and panpsychism is the related notion that every object has a mind.[3] Already my car talks to me, as do my phone, my computer, and my refrigerator, so I guess we could live with talking rocks, chairs, logs, sandwiches, and atoms. And, unlike the chirping electronic appliances, the talking objects might truly have soul.

My opinion is that consciousness is not so very hard to achieve. How does everything wake up? I think the key insight is this:

> ***Consciousness = universal computation + memory + self-reflection***

Computer scientists define universal computers as systems capable of emulating the behavior of every other computing system. The complexity threshold for universal computation is very low. Any desktop computer is a universal computer. A cell phone is a universal computer. A Tinkertoy set or a billiard table can be a universal computer.

In fact, just about any natural phenomenon at all can be regarded as a universal computer: swaying trees, a candle flame, drying mud, flowing water, even a rock. To the human eye, a rock appears not to be doing much. But viewed as a quantum computation, the rock is as lively and seething as, say, a small star. At the atomic level, a rock is like a zillion balls connected by force springs; we know this kind of compound oscillatory system behaves chaotically, and computer science teaches us that chaotic systems can indeed support universal computation.

The self-reflection aspect of a system stems from having a feedback process whereby the system has two levels of self-awareness: first, an image of itself reacting to its environment, and second, an image of itself watching its own reactions.[4]

We can already conceive of how to program self-reflection into digital computers, so I don't think it will be long until we can make them conscious. But digital computers are *not* where the future's at. We don't use clockwork gears in our watches anymore, and we don't make radios out of vacuum tubes. The age of digital computer chips is going to be over and done, if not in a hundred years, then certainly in a thousand. By the Year Million, we'll be well past the Great Awakening, and working with the consciousness of ordinary objects.

I've already said a bit about why natural systems are universal computers. And the self-reflection issue is really just a matter of programming legerdemain. But two other things will be needed.

First, in order to get consciousness in a brook or a swaying tree or a flame or a stone, we'll need a universal memory upgrade that can be, in some sense, plugged into natural objects. *Second*, for us to be able to work with the intelligent objects, we're going to need a strong form of nondigital telepathy for communicating with them.

In the next section, I'll explain how, before we bring about the Great Awakening, we'll first have to manipulate the topology of space to give endless memory to every object, and then create a high-fidelity telepathic connection among all the objects in the world. But for now let's take these conditions for granted. Assume that everything has become conscious, and that we are in telepathic communication with everything in the world.

To discuss the world after this Great Awakening, I need a generic word for an uplifted, awakened natural mind. I'll call these minds *silps.* We'll be generous in our panpsychism, with every size of object supporting a conscious silp, from atoms up to galaxies. Silps can also be found in groupings of objects—here I'm thinking of what animists regard as *genii loci*, or spirits of place.

There seems to be a problem with panpsychism: how do we have synchronization among the collective wills involved in, say, rush-hour traffic? Consider the atoms, the machine parts, the automotive subassemblies, the cars themselves, the minds of the traffic streams, not to mention the minds of the human drivers and the minds of their body cells. Why do the bodies do what the brains want them to? Why do all the little minds agree? Why doesn't the panpsychic world disintegrate into squabbling disorder? Solution: everyone's idea of their motives and decisions are *Just So* stories cobbled together ex post facto to create a narrative for what is in fact a complex, deterministic computation, a lawlike cosmic harmony where each player imagines he or she is improvising.

It takes some effort to imagine a panpsychic world. What would a tree or campfire or waterfall be into? Perhaps they just want to hang out, doing nothing. Perhaps it's only we who want to rush around, fidgety monkeys that we are. But if I overdo the notion of silp mellowness, I end up wondering if it even matters for an object to be conscious. Assuming the silps have telepathy, they do have sensors. But can they change the world? In a sense, yes: if silps are quantum computations, then they can influence their own matter by affecting rates of catalysis, heat flows, quantum collapses, and so on.

Thus a new-style silp drinking glass might be harder to break than an old-style dumb glass. The intelligent, living glass might shed off the vibration phonons in optimal ways to avoid fracture. In a similar connection, I think of a bean that slyly rolls away to avoid being cooked; sometimes objects do seem to hide.

The remarks about the glass and the bean assume that silp-smart objects would *mind* being destroyed. But is this true? Does a log mind being burned? It would be a drag if you had to feel guilty about stoking your fire. But silps aren't really likely to be as bent on self-preservation as humans and animals are. We humans (and animals) have to be averse to death, so that we can live long enough to mate and to raise our young. Biological species go extinct if their individuals don't care about self-preservation. But a log's or rock's individual survival doesn't affect the survival of the race of logs or rocks. So silps needn't be hardwired to fear death.

Let's say a bit more about self-reflection among silps. As a human, I have a mental model of myself watching myself have feelings about events. This is the self-reflection component of consciousness mentioned above. There seems no reason why this mode of thought wouldn't be accessible to objects. Indeed, it might be that there's some "fixed point" aspect of fundamental physics making self-reflection an inevitability. Perhaps, compared to a quantum-computing silp, a human's methods for producing self-awareness is weirdly complex and roundabout.

As I mentioned before, when the Great Awakening comes, the various artificially intelligent agents of the Orphidnet will be ported into silps or into minds made up of silps. As in the Orphidnet, we'll have an upward-mounting hierarchy of silp minds. Individual atoms will have small silp minds, and an extended large object will have a fairly hefty silp mind. And at the top we'll have a truly conscious planetary mind: Gaia. Although there's a sense in which Gaia has been alive all along, after the Great Awakening, she'll be like a talkative, accessible god.

Because the silps will have inherited all the data of the orphids, humans will still have their omnividence, their shared memory access, and their intelligence amplification. I also predict that, when the Great Awakening comes, we'll have an even stronger form of telepathy, which is based upon a use of the subdimensions.

Exploiting the Subdimensions

Let's discuss how we might provide every atom in the universe with a memory upgrade, thus awakening objects to become silps. And, given that the silp era will supersede the nanotech era, we'll also need a nonelectronic form of telepathy that will work after the Orphidnet and digital computers have withered away.

To achieve these two ends, I propose riffing on an old-school science-fiction power chord, the notion of the "subdimensions." The word is a science-fiction shibboleth from the 1930s, but we can retrofit it to stand for the topology of space at scales below the Planck length—that is, below the size scale at which our current notions of physics break down.

One notion, taken from string theory, is that we have a lot of extra dimensions down there, and that most of them are curled into tiny circles. For a mathematician like myself, it's annoying to see the physicists help themselves to higher dimensions and then waste the dimensions by twisting them into tiny coils. It's like seeing someone win a huge lottery and then put every single penny into a stodgy, badly run bond fund.

I recklessly predict that sometime before the Year Million we'll find a way to change the intrinsic topology of space, uncurling one of these stingily rolled-up dimensions. And of course we'll be careful to pick a dimension that's not absolutely essential for the string-theoretic Calabi-Yau manifolds that are supporting the existence of matter and spacetime. Just for the sake of discussion, let's suppose that it's the eighth dimension that we uncurl.

I see our eighth-dimensional coils as springing loose and unrolling to form infinite eighth-dimensional lines. This unfurling will happen at every point of space. Think of a plane with hog bristles growing out of it. That's our enhanced space after the eighth dimension unfurls. And the bristles stretch to infinity.

And now we'll use this handy extra dimension for our universal memory upgrade! We'll suppose that atoms can make tick marks on their eighth dimension, as can people, clouds, or stones. In other words, you can store information as bumps upon the eighth-dimensional hog bristles growing out of your body. The ubiquitous hog bristles provide endless memory at every location, thereby giving people endless perfect memories, and giving objects enough memory to make them conscious as well.

Okay, sweet. Now what about getting telepathy without having to use some kind of radio-signal system? Well, let's suppose that all of the eighth dimensional axes meet at a point at infinity, and that our nimble extradimensional minds can readily traverse an infinite eighth-dimensional expanse, so that a person's attention can rapidly dart out to the shared point at infinity. And once you're focused on the shared point at infinity, your attention can zoom back down to any space location you like.

In other words, everyone is connected via an accessible router point at infinity. So now, even if the silps have eaten the orphids as part of the Great Awakening, we'll all have perfect telepathy.

(Can we travel an infinite distance in a finite time? Perhaps we'll use a Zeno-style acceleration, continually doubling our speed. Thus, traversing the first meter along the eighth dimensional axis might take a half a millisecond, the second meter a quarter of a millisecond, the third meter an eighth of a millisecond, and so on. In this fashion your attention can dart out to infinity in a millisecond.)

The End?

Of course we won't stop at mere telepathy! By the Year Million, we'll have teleportation, telekinesis, and the ability to turn our thoughts into objects.

Teleporting can be done by making yourself uncertain about which of two possible locations you're actually in—and then believing yourself to be "there" instead of "here." We'll work out this uncertainty-based method of teleportation as a three-step process. First, you perfectly visualize your source and target locations and mentally weave them together. Second, you become uncertain about which location you're actually in. And third, you abruptly observe yourself, asking, "Where am I?" Thereby you precipitate a quantum collapse of your wave function, which lands you at your target location. I'm also supposing that whatever I'm wearing or holding will teleport along with me; let's say that I can carry anything up to the weight of, say, a heavy suitcase.

Once people can teleport, they can live anywhere they can find a vacant lot to build on. You can teleport water in and teleport your waste away. What about heat and light? Perhaps you can get trees to produce electricity, and then set sockets into the trunks and plug in your lamps and heaters. Or just get the trees to make light and heat on their own, and never mind the electricity. (Once we can talk to our plants, it should be fairly easy to tweak their genes.)

As the next step beyond teleportation, we'll learn to teleport objects without having to move at all. This long hoped-for power is known to science fiction writers and psi advocates as *telekinesis* or *psychokinesis*. How might telekinesis work in our projected future? Suppose that, sitting in my living room, I want to teleport an apple from my fridge to my coffee table. I visualize the source and target locations just as I do when performing personal teleportation; that is, I visualize the fridge drawer and the tabletop in the living room. But now, rather than doing an uncertainty-followed-by-collapse number on my body, I need to do it on the apple. I become the apple for a moment, I merge with it, I cohere its state function to encourage locational uncertainty, and then I collapse the apple's wave function into the apple-on-table *eigenstate*.

What's the status of the apple's resident silp while I do this? In a sense the silp *is* the apple's wave function, so it must be that I'm bossing around the silp. Fine.

Can animals and objects teleport as well? What a mess that would be! We'd better hope that only humans can teleport. How might we justify such a special and privileged status for our race?

I'll draw on a science-fictional idea in a Robert Sheckley story.[5] Sheckley suggested that humans would have the power of teleportation because, unlike animals or objects, we experience doubt and fear. Certainly it seems as if animals don't have doubt and fear in the same way that we do. If a predator comes, an animal runs away, end of story. If cornered, a rat bares his teeth and fights. Animals don't worry about what *might* happen; they don't brood over what they did in the past; they don't agonize over *possibilities*—or at least one can suppose that they don't.

And it's easy to suppose that the silps that inhabit natural processes don't have doubt and fear either. Silps don't much care if they die. A vortex of air forms and disperses, no problem.

So *why* would doubt and fear lead to teleportation? Having doubt and fear involves creating really good mental models of alternative realities. And being able to create good mental models of alternative realities implies the ability to imagine yourself being *there* rather than *here*. We can spread out our wave functions in ways that other beings can't. Humans carry out certain delicate kinds of quantum computations—which, we can suppose, might lead to teleportation.

Take this to the extreme. Could we create objects out of nothing? Call such objects *tulpas*. In Tibetan Buddhism, a tulpa is a material object or person that an enlightened adept can mentally create—a psychic projection that's as solid as a brick. I think it's entirely possible that, a million years from now, any human could create tulpas. How? You'll psychically reprogram the quantum computations of the atoms around you, causing them to generate de Broglie matter waves converging on a single spot. Rather than being *light* holograms, these will be *matter wave* holograms—that is, physical objects created by computation: your tulpas.

Your thoughts could become objects by coaxing the nearby atoms to generate matter holograms that behave just like normal objects. You could build a house from nothing, turn a stone into bread, transform water into wine (assuming, given such miraculous abilities, you still needed shelter, food, drink), and make flowers bloom from your fingertips.

And then will humans finally be satisfied?

Of course not. By the Year Million, we'll be pushing on past the realm of the finite and into the transfinite realms beyond the worlds of this local universe.

Into the Very Deepest Future

Chapter 12

The Rise and Fall of Time

Sean M. Carroll

Imagine that we prepare a time capsule to be opened one million years in the future. We build it very carefully so that it survives without damage and remains absolutely impregnable to outside influence. Putting aside the question of what we might want to put inside the capsule, what would our future friends expect to find when they open it?

I can tell you one thing they would *not* expect to find inside a million-year-old impregnable capsule of modest size: something still flapping around, gasping for air, wondering what had taken you so long. It's hard to imagine, absolutely sealed inside a tight container, unable to exchange energy or information with the outside world, an organism or a device that would manage to keep going over the course of a million years. Something could be set into motion by the act of opening the box (if a switch were flipped, or two chemicals mixed together), but nothing is going to squirm around continuously for that long. This isn't a statement about our technological capability; it's a general fact about physical systems. Left alone, systems tend to wear themselves out, relaxing into some static, stable configuration, and then just sitting there—forever.

Winding Down

This tendency toward stasis isn't always obvious to us, because usually we deal with objects that are not isolated from the rest of the universe. We see things moving all around, showing no signs of slowing down. It's true that living organisms tend to die, but even then motion doesn't cease entirely; corpses decompose, get consumed by microorganisms, and their useful molecules returned to the biosphere as part of the cycle of life.

But the cycle of life is not a perpetual motion machine. We're fortunate that the Sun pumps copious amounts of energy to us. If that source were magically to disappear, then life on Earth would grind to a halt in a relatively short time. And it's not just that we get energy from the Sun; we get *useful* energy, in the form of concentrated visible light, which can be reradiated at lower energies into the darkness of space after we've put it to work. If we were surrounded by suns on all sides, Earth would reach the temperature of the solar photosphere, and life would die off just as quickly and surely as if the Sun were to disappear entirely.

Well, you say, there's more to the universe than life. In the nonbiological realm, we observe the circulation of water through the atmosphere, the re-arrangement of tectonic plates, and the stately orbits of planets and comets through the solar system. These seem to be forms of motion that could well continue forever.

But they won't. The circulation of water, of course, is driven by the same solar energy that powers life. The motion of the continents will eventually grind to a halt as Earth's core cools, once its radioactive elements all decay away.

And even the solar system is not forever. Planetary orbits seem stately, but they're actually chaotic; if we wait long enough, some planets will spiral into the Sun, while others will be ejected into interstellar space.

And then we have the whole shebang itself. Our universe is in motion, but it's winding down. Our far future is bleak and cold. Stars are using up their nuclear fuel, turning lighter elements into heavier ones. Galaxies are moving apart, their light ever more red and faint. We used to imagine that the universe would someday hit a maximum size, then start contracting into a noisy Big Crunch, which at least would be exciting while it lasted. But recent cosmological observations indicate that our universe is not only expanding, but *accelerating*—the rate at which any particular galaxy is moving away from us is going up, not down. This sounds like a form of sustainable motion—and it is—but it's not the kind of interesting, complicated evolution that characterizes biology. Everything is moving apart, cooling off, slowing down. Eventually the universe will be a collection of black holes, dark planets, and elementary particles; ultimately even the black holes and lightless planets will dissipate, leaving nothing but a thin gruel of particles ever more dispersed into the featureless void. And it will be like that, as far as we know, forever.

The eventual "heat death" of the universe is a far-off prospect—well more than a million years in the future, well more than a billion. (The somewhat misleading term *heat death* does not mean "death by heat," of course, but rather the death of heat itself.) But if we want to have any hope of predicting what things will be like in the relatively near future, then we should try to understand the underlying properties of time and motion. Why, if the universe is winding down, hasn't it done so already? And what does that teach us about our own future?

Cosmologists and physicists have assembled a laundry list of features of the physical world that make us furrow our brows and ask "Why is it like that?" But the very fact that matter in the cosmos is still moving, rather than at rest, is the single most surprising thing about the universe.

History of the Universe

The fact that our universe is unnaturally lively is an under-appreciated problem, even by professional cosmologists, so let's drive it home more explicitly. We can start with the end (of the universe), and then go back to the beginning.

Theorists up to and including Albert Einstein argued that a single episode of expansion and then recollapse was preferable to one in which expansion continues forever. A universe that is compact in both space and time seems somehow more manageable and comprehensible—not to mention, more symmetric—than one just growing without bound. But sometimes reality chooses not to conform to our preferences. Multiple lines of astronomical evidence—from supernovas, the cosmic microwave background, large-scale structure, gravitational lensing, and more—have recently convinced cosmologists that the expansion of the universe is accelerating rather than slowing down. The simplest explanation for such behavior is the existence of dark energy: a smooth, persistent component of energy that is spread uniformly throughout space and imparts an undiluted impulse to expand. There are a number of candidates on the market for what the dark energy might be, but the simplest is *vacuum energy*, or the cosmological constant—an absolutely constant trace of energy inherent in every cubic centimeter of space, whether it's bubbling with particles or utterly empty.

The dark energy that makes the universe accelerate could be temporary and might someday fade away. But the most straightforward interpretation, which fits all of our current data perfectly, is that the dark energy is constant and will stay that way. If so, there's no prospect of the universe recollapsing. It will continue to expand, matter and radiation will die away, the cycles of stellar and galactic evolution will use up their fuel and give out, and we'll be left with a cold, empty cosmos. Every particle will be moving away from every other particle, and eventually these particles won't even be able to see each other. For all intents and purposes, motion will have ceased forever.

And we should emphasize *forever*. In this scenario (which seems to be where the smart money is betting these days), the future of the universe stretches into eternity, and it remains cold and empty for an infinitely long time. We've lived in a relatively active cosmos for fourteen billion years, and we have a trillion or so years of life left in us; but all of those big numbers are still *finite*.

In other words: the fraction of the universe's life span during which it is interesting and active is, strictly speaking, zero. A finite post–Big Bang spurt of life, followed by an infinite stretch of lifeless equilibrium. If we were to pick a random moment in the history of the universe, the chance that it would be happily located in the current warm-and-fuzzy phase is literally infinitesimal. And yet here we are. But then, where else could we be?

Departing from Equilibrium

There is a convenient explanation for our apparent good fortune. Interesting dynamical processes occur only during this fleeting fraction of the lifetime of the universe, and we human beings are examples of these processes. We can exist only during the brief period before the matter in the universe settles down to equilibrium. Attempts to explain features of our observable universe by appealing to the fact that its properties *must* allow for our existence belong under the rubric of the "anthropic principle."

The anthropic principle gets a bad rap in certain circles, but there is a relatively innocent (and practically tautological) version: given a variety of possible environments, living organisms will be found only in those conditions that are hospitable to their existence. Most of the volume of space in the solar system is desolate near-vacuum, and yet here we are in the atmosphere of a planet. Should we be surprised? Of course not, since not all parts of the solar system are equally hospitable to life. Given that the infinite future of our universe is even more desolate and inhospitable than the space between our planets, we shouldn't be surprised to find ourselves in the early hospitable period of the universe's history.

The question, rather, is why there is such an early hospitable period at all. Why hasn't the universe always been in a cold, empty state of equilibrium? This question was taken seriously by Ludwig Boltzmann, one of the founders of modern statistical mechanics. Working in the late nineteenth century, Boltzmann was the first to grasp the true meaning of the concept of entropy. Before him, pioneers in the field of thermodynamics—many of them practical folk, interested in building better steam engines and refrigerators—understood entropy as a measure of the *uselessness* of a certain amount of energy. There is energy in the pressurized gas contained in a sealed balloon—we can let the gas escape and use its motion to do work by turning a flywheel. Once the gas mixes with the surrounding air, there's nothing more we can do with it; the total energy is the same, but the entropy is now higher.

Boltzmann had the crucial insight that entropy counts the number of indistinguishable ways we can arrange things. For a gas, we can rearrange individual molecules without changing the gross features of the system (density, temperature, pressure). When a certain amount of gas is in the balloon, it's clearly segregated from the surrounding air, and there are only so many arrangements available to us. Once it has escaped, the gas molecules mix thoroughly with the surrounding air, and many more arrangements become possible.

Boltzmann's reimagining of entropy as the number of indistinguishable ways we can arrange the components of the system sheds new light on the second law of thermodynamics: the entropy of a closed system will never spontaneously decrease. Formerly interesting but mysterious, it now makes perfect sense. Of course entropy tends to go up, because there are many more ways for a system to have high entropy than low entropy!

Sure, Boltzmann's understanding of entropy explains

why low-entropy states tend to evolve into high-entropy states. But that's not quite enough to account for why we see the entropy changing all around us. We need to explain one more thing: why was the entropy ever small in the first place?

Thermal Fluctuations

Boltzmann understood this problem. He also knew that his novel understanding of entropy provided a possible answer. From the Boltzmannian perspective, entropy tends to go up just because it's overwhelmingly likely to do so, since there are many more high-entropy states than low-entropy ones. But it's not *necessary* that the entropy increase; it's just very probable. The second law of thermodynamics is a *statistical* rule, not an absolute one. If we wait long enough, very rare events will eventually occur, including spontaneous decreases in entropy. (Indeed, small-scale decreases of entropy have been observed in experiments.)

Could this statistical logic explain the universe? Maybe, from an ultra-big-picture perspective, the universe *is* in thermal equilibrium. At almost all times, it is cold and empty and unchanging. But an occasional fluctuation (or so the story would go) throws it out of equilibrium, into something that looks like our current universe, and that's where life can be found.

Sadly, this story doesn't quite hold together. Fluctuations in entropy are rare, but more importantly, big fluctuations are far more rare than small fluctuations. It's unlikely, but possible, for all of the air molecules to collect briefly on one side of the room; but it's even more unlikely that they all collect in a tiny region in the corner. And our universe has a *much* lower entropy than it would if it were in equilibrium. So again, why isn't it?

Physicists still don't understand the entropy of the gravitational field well enough to put exact numbers on such quantities, but we can make reasonable estimates. The matter within our current universe has an entropy of about 10^{100}. (The largest single contributor to this number is the entropy of supermassive black holes at the center of galaxies.) At early times, when matter was smoothly distributed and there were no black holes, the corresponding entropy was about 10^{88}. And in a maximum-entropy configuration, it would be about 10^{120}.

Those are all big numbers, but—and this may not be obvious—some are *vastly* bigger than others. When you subtract 10^{100} from 10^{120}, the result is basically still 10^{120}. (Just as when you subtract one from a million, what you're left with is basically a million.) According to Boltzmann, thermodynamic systems can randomly fluctuate to configurations of lower entropy, but the probability of doing so is given by $e^{-\Delta S}$, where ΔS (pronounced "delta-S") is the change in entropy between the two configurations. So the chance that the stuff of which our universe is made underwent a fluctuation from equilibrium to its present configuration is about $\exp(-10^{120})$. A tiny number, by any measure.

But given an infinitely long wait, eventually it would happen. That's not the problem with this way of explaining the present state of the universe. The problem is that the early universe had an even *lower* entropy! Even if we think that we couldn't exist unless the entropy of the universe were as low as it is today (which seems a bit self-centered of us, but anyway), there's certainly no good reason for it to have been even lower in the past. The second law of thermodynamics makes sense to us because the early universe had a low entropy; but our very existence does not depend on it.

What We See—and Don't

So we can conclude that our observed universe is not a temporary fluctuation around an equilibrium configuration. If it were, then there would be no reason for the fluctuation to have been so dramatic. Our universe has a much lower entropy than there is any reason for it to have.

The usual explanation for the strange temporal imbalance between the hot past back to the Big Bang and the cold infinite future is—well, there isn't any usual explanation. It's an unsolved problem within the framework of modern cosmology. To many researchers, however, it doesn't seem an extremely pressing problem, for the simple reason that the Big Bang is mysterious. In classical general relativity, the Bang is a singular moment of infinite density and spacetime curvature. But nobody expects classical general relativity to be the final answer. General relativity isn't compatible with the postulates of quantum mechanics, which govern the behavior of the microscopic world. To really understand the extreme conditions near the Big Bang, we will need a quantum theory of gravity. Since such a theory remains elusive (string theory is our best bet, but far from complete), many people assume that what really happened near the Big Bang is the kind of question we are simply not equipped to address yet.

And they may be right. But brushing the puzzle of the early universe's very special conditions under the rug of the Big Bang doesn't seem quite fair. The dramatic asymmetry between the early and late phases of the universe isn't simply an academic cosmological curiosity; it's something that shows up right here and now in everyday manifestations of the asymmetry between past and future. Our local environment exhibits a strong "arrow of time"—ice cubes melt in water but don't spontaneously appear in glasses; it's easy to turn eggs into omelets but impossible to get them back again; we can remember what happened in the past but we certainly don't remember the future. And all of this can be traced to special low-entropy conditions in the early universe. But it's not honest to say "we explain the arrow of time by noting that the initial conditions are different from the final conditions." That's not *explaining* temporal asymmetry at all; that's just asserting it baselessly.

The truth is: nothing we understand about physics within our observed universe suffices to explain the arrow of time. Perhaps the key word here is "observed." Increasingly, cosmologists are willing to entertain the idea that there is a lot of universe out there that we *don't* observe. We can imagine that the Big Bang wasn't actually the beginning of everything, and that space and time both existed even before the Bang. And we can also imagine that there are different physical locations far away from us in space, which we simply don't see because the limiting speed of light prevents information from traveling so far so soon. These hypothetical regions, earlier in time or farther away in space than we are able to see, together constitute what is sometimes called the *multiverse*.

It can't be stressed enough that the multiverse is utterly hypothetical—but that doesn't mean it isn't real.

How does a multiverse help us with the arrow of time? It sidesteps the problem of explaining the special low-entropy initial conditions by suggesting that those conditions aren't really "initial." If there was time before the Big Bang, then our universe came from somewhere; there was some physical process in the preexisting multiverse that gave rise to the conditions that seem "early" to us. Now the question has changed from "Why did the universe start in such a finely tuned state?" to "What kind of process would create such a finely tuned state?" Even if we don't know the answer, a change of perspective can help us improve the question.

Entropy and Complexity

There is another curious feature of the evolution of our universe. At one end of time—the beginning—the universe is in a special state, with an incredibly low entropy: it's a hot, dense plasma, smoothly distributed throughout space, with small perturbations of about equal amplitude on all length scales. At the other end of time—the very far future—the universe is in a much more typical-looking state, with high entropy: it's cold, empty space, with a small vacuum energy. But at both the beginning and the end, a *simple* description suffices to capture the essential features of the universe (as we just did).

For a brief slice of time between the two, however, including the present time, the state of the universe is *complex*. No simple one-line description could possibly serve to capture what is currently going on. There is ordinary matter and dark matter, and the ordinary matter comes in a bewildering variety of ions, atoms, and molecules in different forms and combinations. Matter is grouped into clouds, planets, and stars, grouped into galaxies of various sorts, which in turn belong to clusters and superclusters. Each star and planet is the scene of a great deal of intricate nuclear and chemical reactions, all the way up to the ultimate in complex reactions: organic life. To capture the important features of the present universe, we must not only describe all of the different forms in which matter is arranged, but also specify the relevant location of *each individual piece* of complex matter. (A location is "relevant" if changing it would substantially alter the dynamics of the object in question. We can exchange the position of two rocks without its mattering too much; but switching one rock with a person's head is, in most cases, not likely to go unnoticed.)

This complexity developed gradually over time, from the early universe to today, and it will gradually fade away as the universe grows older. In the short term, we can exercise and eat right in an attempt to fend off the ravages of time; on cosmological scales, however, our protons will probably decay and our elementary constituents will be scattered throughout the universe, possibly passing through a black hole in the process.

So the evolution of entropy as time passes is straightforward: it keeps going up. But the evolution of *complexity* is, for want of a better term, more complex. It rises at first, before declining and giving way to simplicity. But the two phenomena are nevertheless closely connected. In a real sense, the complexity of form and interaction that is familiar in our everyday lives characterizes an ephemeral stage in the history of the universe; it rides parasitically on the change of entropy with the passage of time. Everything interesting in our lives, from the formation of memories to the creation of works of art, is a process that increases the total entropy of the universe. But that's only possible in a universe where the total entropy is changing. The anthropic demand that the universe we observe be hospitable to life doesn't suggest simply that the entropy is not at its maximal value; it also implies that it's not remaining constant at any lower value. We're complex beings, and a simple universe can't sustain us.

Hierarchy of Scales

The growth of entropy from a low value in the early universe to a high value in the late universe is a statement about the configuration in which we find the physical system we call "the universe." But the rich complexity we've just discussed also relies on a special property of the dynamical laws governing the universe's evolution: the wide separation of different energy scales that arise as fundamental parameters.

The reason why planets and stars and galaxies have the sizes they do can be traced back to the fundamental interactions of physics: electromagnetism, gravity, and the strong and weak nuclear forces. Each of these is associated with an energy scale. Unless you're a physicist, the following equations probably won't mean anything to you. Luckily, the details of these energy scales and their definitions are not important for our discussion. What's important to see is that they are all very *different*, ranging over twenty-six orders of magnitude, where each order of magnitude is ten times larger than the one beneath it.

For gravity, the relevant energy scale is the Planck energy, constructed by combining Newton's gravitational constant with Planck's constant (from quantum mechanics) and the speed of light: $E_p = \left(\frac{\hbar c^5}{G}\right)^{\frac{1}{2}} = 10^{27}$ eV (where eV stands for "electron volt," a common unit of energy in particle physics). The weak force is carried by the *W* and *Z* bosons, which have masses of about 8×10^{11} eV. The strong force, described by QCD (quantum chromodynamics), is associated with the QCD scale, $\Lambda_{QCD} = 3 \times 10^8$ eV. Finally, electromagnetism is characterized by the typical binding energy of an electron to an atomic nucleus; that's given by the Rydberg, around 10 eV. We could also toss in the vacuum energy, which is associated with a scale of about one thousandth of an electron volt.

Physicists like to say that there is a *hierarchy* among the fundamental scales of elementary particles. With the exception of the QCD scale, we currently have no reliable understanding of why such a hierarchy exist. If you didn't know any better, the rules of quantum field theory would lead you to guess that all of these scales should be close to each other. Attempts to solve the puzzle of this wide separation of scales—especially the differences between the vacuum energy scale and the Planck scale, known as the "cosmological constant problem," and between the weak scale and the Planck scale, known simply as "the hierarchy problem"—are the focus of intense effort in contemporary physics.

In the meantime, we can all be thankful that the fundamental scales of physics *are* widely separated, even if we can't explain why. Just as complexity depends on the growth of entropy, it also depends on the hierarchy of fundamental scales. One reflection of that complexity is the difference in regimes in which different forces of nature dominate. On the scale of atomic nuclei, the strong nuclear force (and to a lesser extent the weak force) is most important. But, while the strong force is, well, strong, it is also limited in range. Over larger distances, electromagnetism takes over; indeed, almost all of chemistry and the physics of materials can ultimately be understood in terms of electromagnetic interactions. However, electric charges come in both positive and negative forms, and so tend to cancel each other into neutrality for sufficiently large objects.

On the very largest scale, gravity dominates. The weakest of the forces, it stretches over a long range and always accumulates rather than canceling out, since there's no such thing as a negative gravitational charge. (We can imagine exotic matter with negative mass that would be gravitationally repulsive. Actually, some theorists of wormhole travel propose using such stuff as struts to hold the wormholes open against crushing gravity, but it's worse than hypothetical; there are very good reasons to believe that it can't exist. Nor does dark energy count as "antigravity" even though it's shoving the galaxies apart; its energy density is positive, even if it exerts negative pressure.)

This story should sound familiar—we find something unnatural about the universe we observe, but it's something on which our very existence depends. That holds true for the arrow of time and the evolution of our cosmological configuration from a low-entropy state to a high-entropy one, as well as for the dramatic hierarchy separating the fundamental scales of particle physics. It's tempting to wonder whether both kinds of puzzle can be resolved by imagining that our observable part of spacetime is embedded in a larger multiverse.

The answer seems to be: very possibly. (We don't know yet, of course, but it would help to explain the conditions of our universe and, beyond that, is a natural extension of the Copernican principle: our universe should hold no more privileged a place within the greater scheme than Earth holds within our universe.) Instead of going back in time to before the Big Bang, what we contemplate now is a collection of different regions of *space*, each with different physical parameters. And that's exactly what we seem to get out of string theory, our leading candidate for a quantum theory of gravity. String theory doesn't uniquely predict the laws of physics as we observe them. Rather, it predicts a "landscape" of different possible versions of low-energy physics, each of which could plausibly exist somewhere in the wider universe. Maybe the reason why we find such a rich hierarchy within the laws of physics is that some parts of the universe exhibit them and some don't—but observers such as ourselves pop up to think about the issue in only those interesting regions where the hierarchy is substantial.

Taking the Multiverse Seriously

Let's recap the reasoning that has led us to this point. When we think about isolated physical systems, we notice that they tend to wind down, evolving to a state of maximum entropy and thereafter remaining in equilibrium. We might have expected the universe itself to have reached such an equilibrium long ago, but such is far from the case; rather, the early universe was in such an extremely low-entropy state that we have a long way to go before we reach anything resembling equilibrium. This interesting state of affairs can be traced to special conditions at the Big Bang, raising the possibility that such conditions could have characterized a universe that existed before the Big Bang. Similarly, the laws of physics appear to feature a variety of widely separated scales, in contrast to what we might naively imagine. This raises the possibility that there are different regions of space, with different sets of physical parameters, but that we can exist only when they are arranged into the appropriate hierarchy.

So "unnatural" features of our observable universe might reflect its ultimate status as part of a much bigger multiverse, one in which our local environment is far from typical. Nobody can deny the speculative nature of this idea, but we are driven to contemplate it by an urge to understand features that we observe here and now. Unfortunately, simply chanting "multiverse" does not resolve all of our puzzles at once. If the special conditions at the Big Bang arise from its origin as a baby universe born from a parent spacetime, then it remains to be understood why the particular conditions we observe are favored. If the observed hierarchy of particle physics is necessary for the existence of complex creatures such as ourselves, then we still need to describe the set of all possible laws of physics and how they are distributed through space. The multiverse idea has a long way to go before it becomes a respectable scientific theory.

But that doesn't mean that we can't indulge our imaginations a bit. Taking together the lessons of quantum mechanics and general relativity, we can contemplate the possibility that an appropriate collection of energy can rip spacetime apart and give rise to a baby universe. The newborn universe features a certain specific configuration of matter in a low-entropy state, and some specific parameters governing the local laws of physics. Such an arrangement can arise, we imagine, spontaneously, as the outcome of quantum fluctuations that are individually quite rare, but ultimately inevitable when considered over an infinite stretch of cosmic time.

But if something can happen naturally, perhaps it can also be brought about artificially. If Nature can give birth to a baby universe, then can we help it along? Can we collect the right combination of fields in the appropriate configuration to induce the pinching off of a completely new piece of spacetime? Even better, could we fine-tune our engineering so that the consequent universe had some particular set of observable physics, specified by us [as described by Amara Angelica in Chapter 10, where prior intelligent life "tunes" subsequent universes to be life-friendly—Ed.]? In effect, we would become the God of Newton and the Deists, who set the universe in motion and was strictly noninterfering thereafter.

It's a wild notion, and current wisdom would strongly advise betting against the possibility that universe creation will ever become a commonplace event. But if the laws of physics turn out to allow it, then it's ultimately an engineering problem, and with time even the most challenging engineering problems tend to be solved. Perhaps by the Year Million, our descendants will be creating entire universes. After all, a million years is, by human standards, a very long time indeed.

Chapter 13

The Final Dark

Gregory Benford

> [T]he use, however haltingly, of our imaginations upon the possibilities of the future is a valuable spiritual exercise.
>
> —J. B. S. Haldane, 1923

How it all began, the universe, is a very old obsession. Less often fretted over is the symmetric question: how will it all end?

Robert Frost's famous imagery—fire or ice, take your pick—pretty much sums it up. But lately, largely unnoticed, a revolution has unwound in the thinking about such matters, in the hands of that most rarefied of tribes, the theoretical physicists. Maybe, just maybe, ice isn't going to be the whole story.

Of course, linking the human prospect to cosmology itself is not at all new. The endings of stories are important to us because we believe that how things turn out implies what they ultimately mean. This comes from being pointed toward the future, as any ambitious species must be. There are three forms of chimpanzees: the common chimp, the bonobo, and us. We are the only chimp who got out of Africa. That experience reflects and probably laid down the deep human urge—indeed, our signature: the restless urge to move on, explore, exploit. Natural selection gives us a gut imperative that plays out physically and culturally, in pursuit of our goal: the expansion of human horizons.

On Earth, horizons sufficed for many millennia. But that time is over; the skies beckon, and it is natural to think in terms of our horizons in time. We have cosmology to aid us now, unlike people only a century ago. Most of us believe that physics can tell us more about our prospects than religion. Still, we do think long, and often with theological implications. The far future matters for very basic reasons.

Our yearning for connection explains many cultures' ancestor worship: we enter into a sense of progression, expecting to be included eventually in the company. Deep within us lies a need for continuity of the human enterprise, perhaps to offset our own mortality. Deep time in its panoramas, both past and future, redeems this lack of meaning, rendering the human prospect again large and portentous. We gain stature alongside such immensities. But this flattering perspective sets an ultimate question: will a time come when humanity itself will not be remembered, our works lost and gone for nothing?[1]

A major change in our ideas of cosmology occurred only a decade ago [as discussed in Sean Carroll's chapter—Ed.], with the discovery that the expansion of our universe is accelerating. To reach such an astonishing conclusion demanded new measurements of supernova brightness in faraway galaxies, meanwhile eliminating many possible sources of error, combined with precise calibrations of their distances. Together, these showed that the farther away, the faster others were fleeing from us, and we from them, as we share a quickening expansion. This new finding, that spacetime is opening ever faster, relies upon fairly tricky measurements. It is not easy to study whether the momentary luminosities of supernovas in very distant galaxies fit a pattern. It remains to be extensively checked, but suppose for the moment we take it as given. What follows?

Acceleration implies an ever-bigger cosmos. Some feel repulsed by the entire notion. Cyclic universes have great appeal, as every public lecturer on cosmology knows from the audience questions. Evolution may have geared us to expect cycles; the days and seasons deeply embedded this in our ancestors. The ancient Hindu system embraces it especially, holding that we are already uncountably far into the oscillations, and the universe is unknowably old.

Love of cyclic universes may come from a deep unease with linear time, one that predates our modern ideas. At least the periodic supplies some rhythm, a pattern, rolling hills rather than just a flat plain stretching to infinity. This feeling finds an echo in other common audience questions. *Doesn't all this have a purpose, an end? Does the drama go on forever, really?* But then, genuinely endless repetition also seems to revolt most of the cyclic devotees. They still want to avoid the abyss of infinite time. The Hindu timescale is immeasurably long but not infinite. Many are horrified by a universe that lasts only a finite time, ending in cold or heat. Even placing the event in the very far future, long after our personal deaths, carries the heavy freight of making what we do now meaningless because it does not last. Recall the scene in *Annie Hall* when young Woody Allen refuses to do his homework because the universe is going to end anyway.

Will Shakespeare endure literally forever? As Bertrand Russell put it in *Why I Am Not a Christian*

> All the labours of the ages, all the devotion, all the inspiration, all the noonday brightness of human genius are destined to extinction in the vast heat death of the solar system, and . . . the whole temple of man's achievement must inevitably be buried beneath the debris of a universe in ruins.

So Russell doesn't believe in God because nothing lasts. At first this seems an odd argument, but it goes to our deep questions. If nothing lasts, what is our purpose? Some fervent believers attack the second law of thermodynamics (the heat death) for exactly this reason. Ironically, these Christians join company with atheist Friedrich Engels, who disliked entropy because it would destroy historical progress in the long run.

Suppose we could create a heaven on Earth, or at least somewhere, by the Year Million. Permanent, unchanging paradise seems boring to many, at least if it means mere joyful indolence. Is perpetual novelty even possible, though? Can we think an infinite variety of thoughts?

In 1979 the celebrated Princeton physicist Freeman Dyson brought this entire issue to center stage for physicists and astronomers. He already had his prejudices: he wouldn't countenance the Big Crunch option because it gave him "a feeling of claustrophobia." Still, must all our revelries end? Science, he thought, might be able to settle whether a Last Day is ever going to arrive.

When physicists ask questions, they do a calculation to clarify matters. He discussed the prognosis for intelligent life. Even after stars have died, Dyson asked, might life be able to survive forever without intellectual burnout?

Energy reserves will be finite, and at first sight this might seem to be a basic restriction. But Dyson showed that this constraint was not fatal. He looked beyond the time when stars would have tunneled into black holes, which would then evaporate, by comparison, almost instantaneously. As J. D. Bernal foresaw in *The World, the Flesh, and the Devil* (1929):

> Consciousness itself may end . . . becoming masses of atoms in space communicating by radiation, and ultimately resolving itself entirely into light . . . these beings . . . each utilizing the bare minimum of energy . . . spreading themselves over immense areas and periods of time . . . the scene of life would be . . . the cold emptiness of space.

Dyson's answer was positive. He thought that by hibernating, life could endure eternally. But in the third of a century since Dyson's article appeared, our perspective has changed in two ways, and both make the outlook more dismal.

First, Dyson originally assumed matter would last for eternity. Most physicists now strongly suspect that atoms don't live forever. Though the proton lifetime remains unmeasured, current particle theory predicts protons should decay in about 10^{34} years. The basic building block, the proton, will decay into lesser particles. White dwarfs and neutron stars will erode away in about 10^{36} years, sputtering into wan energies and small sprays of electrons and positrons. The heat generated by particle decay will make each star glow, but only as dimly as a domestic heater—no real help against the pervasive cold.

Our universe is about fourteen billion years old, that is, a little over 10^{10} years. In principle, everybody agrees that despite the steady cooling, order could persist even up to 10^{34} years. Here we speak of *unimaginably* long times—except that physicists *have* imagined them, guided by the gliding calculus of theoretical physics. But writing down numbers is a dry way of gaining what we really mean by imagining—that is, having a gut feeling. Still, calculation is all we have to go on.

After protons fade away in, say, 10^{34} years, our local group of galaxies will be just a swarm of dark matter, electrons, and positrons. Thoughts and memories could survive beyond the first 10^{36} years, if downloaded into complicated circuits and magnetic fields in clouds of electrons and positrons—maybe something that will resemble the threatening alien intelligence in *The Black Cloud,* the first and most imaginative of astronomer Fred Hoyle's novels, written in the 1950s.

Dyson felt this would be an "austere mode of existence," and continued, with classic understatement, "even if this assumption is wrong, it is certainly good for the next 10^{34} years, long enough for life to study the situation carefully."

The second bit of bad news is that the accelerating expansion means the universe cools even faster. There is less time to avert the cold—and less room, too.

Why less? Characteristically, Dyson was optimistic about the potentiality of an expanding universe because there seemed to be no limit to the scale of artifacts that could eventually be built. He envisioned the observable universe growing ever vaster. Many galaxies whose light hasn't yet had time to reach us would eventually come into view, and therefore within range of possible communication and networking. Interactions will matter. We could gain knowledge from distant brethren, for use against the encroaching night.

But an accelerating expansion yields a more constricted long-term future. Galaxies will fade from view ever faster as they get more and more red-shifted—their clocks, as viewed by us, will seem to run slower and slower. Then they will seem to freeze at a definite instant, so that even though they never finally disappear we would see only a finite stretch of their future. This is analogous to what happens if a cosmologist falls into a black hole: from a vantage point safely outside the hole, we would see our infalling colleague freeze at a particular time. We'll have only a last snapshot, even though they experience, beyond the horizon, a future that is unobservable to us.

Well before 10^{34} years, our own galaxy, its identical twin neighbor Andromeda, and the few dozen small satellite galaxies that are in the gravitational grip of one or other of them, will merge together into a single amorphous system of aging stars and dark matter. Then the universe will look ever more like an "island system" (the kind of universe originally proposed by Laplace). In an accelerating universe, everything else disappears beyond our horizon. If the acceleration is fixed, this horizon never gets much farther away than it is today.

So there's a firm limit—though, of course, a colossally large one—to how large any network or artifact can ever become. This translates into a definite limit on how complex anything can get. Still worse, one important recent development has been to quantify this complexity limit. Space and time cannot be infinitely divided. The inherent quantum "graininess" of space sets a limit to the intricacy that can be woven into a universe of fixed size. Life has to work within boundaries.

Even if the problem of limited energy reserves could be surmounted—a tall order in itself and the main issue Dyson addressed—there would be a limit to variety and complexity. The best hope of staving off boredom in such a universe would be to construct a time machine and, subjectively at least, exhaust all potentialities by repeatedly traversing a closed time loop. This appears to be possible, within general relativity, but Dyson and others found it also claustrophobic.

But there is other theoretical hope, too. It is a bit abstract, though. Kurt Gödel's famous theorem showed that mathematics contains inexhaustible novelty. That is, there will always be true theorems that can't be proved with what has come before. Only by expanding the conceptual system can they be shown to be true, in a larger view.

Most people would not turn to mathematics for a message of hope, but there it is.

As this darkened universe expands and cools, lower-energy quanta (or, equivalently, radiation at longer and longer wavelengths) can store or transmit information. Just as an infinite series can have a finite sum (for instance, $1 + \frac{1}{2} + \frac{1}{4} + \ldots = 2$), there is perhaps, in principle, no limit to the amount of information processing that could be achieved with a finite expenditure of energy. Any conceivable form of life would have to keep ever cooler, think more slowly, and hibernate for ever-longer periods.

But there would be time to think every thought, Dyson believed, even in the face of the heat death. As Woody Allen once said, "Eternity is very long, especially toward the end."

Life that keeps its temperature fixed will not make it, though. Eventually it will exhaust its energy store. The secret of survival will be to cool down as the universe cools. Being frugal means you could dole out in ever-smaller amounts the energy necessary to live and think.

Silicon or even dust could form the physical basis of such enduring life, at least until the protons decay. After that, there is no fundamental reason why information cannot be lodged in electron-positron plasmas, or even atoms made from them. "Positronium" is an "atom" comprising a positron, or anti-electron, orbiting with an electron, much like a hydrogen atom. In September 2002, a European group succeeded in producing tens of thousands of them in a magnetic bottle, so they could conceivably be used to build solid structures of a wholly new sort.

No matter what the basis of their life is, the crucial distinction for far-future thinkers is their method of storing information. In our computer-saturated world, using information defines life—active flow, not mere passive storage. Life tends to be defined in terms of the reigning paradigm of the time, so in our computer age we make a crucial distinction. There are two choices: analog or digital.

Old-fashioned LP audio records are analog; CDs and

DVDs are digital. Cosmologist Fred Hoyle's ominous *Black Cloud* was analog, storing its memories in magnetic fields and dust particles. A human mind uploaded into a computer would be digital life. Are our brains analog or digital? We do not know yet. But this point became the battleground between Dyson and a bevy of physicists, including Larry Krauss of Case Western University. His challenge to Dyson had the quality of a young buck butting heads with an aging bull. The debate got rarefied right away, including lengthy calculations on the thermodynamics of ultracold, with quantum mechanics for dessert.

Our genetic information carried in DNA is clearly digital, coded in a four-letter alphabet. But the active information in our brains remains mysterious. Memories live in the strengths of synaptic connections between billions of neurons, but we do not fathom how these strengths are laid down or varied. Perhaps memory is partly digital and partly analog; there is no reason why they cannot blend. If we are partly analog, then perhaps the hope of the brain-copying or mind-uploading method will be only partly fulfilled, and some of our more fine-grade thoughts and feelings will not make it into a digital representation.

Actually, the analog/digital divide may not be the whole story. Some theorists think the brain may be a quantum computer, keeping information in quantized states of atoms. But since we know little about quantum computers beyond what's merely possible, the argument over in-principle methods has fastened upon analog versus digital. Interestingly, the long-term prospects of digital intelligences are not the same as analog forms. Krauss leaned heavily on a digital determinism, which shaded quickly into pessimism. Dyson stood his analog ground.

That there is any contest at all may surprise some, since we are so used to analog tools like slide rulers giving way to digital ones like hand calculators. The essential difference is that analog methods deal with continuous variables while digital ones use discrete counting. Surprisingly, analog wins, digital loses. It turns out that the laws of physics allow a thrifty, energy-hoarding information machine (life) to persist, but not a digital one.

The reasons are fairly arcane, involving the quantum theory of information storage. Still, one can think of a digital system as having ratchets that, once kicked forward, cannot go back. As the universe cools, you eventually can't kick the ratchet far enough forward. But a smooth system can inch up as much as you like, storing memories in smaller and smaller increments of energy.

Life can use hibernation to extend its analog form indefinitely. Like bears, it can adapt to falling temperatures by sleeping for progressively longer cosmic naps. Awake, it spends its energy reserves at unsustainable levels. Asleep, it accumulates. It turns out, further, that such life can communicate with other minds over the great distances between galaxies, too. Energy reserves can dwindle, but so does the noise background in the universe, as expansion cools the night sky.

Communication depends not on signal strength (energy) but on the ratio of signal to noise. A cold, expanding universe is friendly to the growth of intergalactic networks. Life will have ample time to wait for an answer from, say, the Andromeda galaxy, without worrying about being able to hear the reply.

But not all is well for analog life if the universe continues to accelerate forever. At some distance, the repulsive force that causes this acceleration must win out over gravity's attraction. So galaxies farther away than this critical distance will accelerate beyond view, setting the limit on the size of structures that life can build. This ultimately dooms it. To persist forever, life needs to be analog, and the universe must not be accelerating forever. The first is an engineering requirement, and presumably savvy lifeforms will heed it. The second we can do nothing about, unless somehow life can alter the very cosmological nature of our universe—surely a tall order.

We do not yet know (and might not for quite a while) whether the acceleration will slow down, because we do not know its cause. This is the biggest riddle in cosmology, and many are pursuing it. The Dyson-Krauss dispute still rages in the hallowed pages of *Physical Review*. Dyson's own vaguely optimistic theology clashes with Krauss's apparent atheism. They are reprising an ancient difference in tastes over the deepest issue: is there any discernible purpose to the universe? And does human action mean anything on this vast stage?

We can't be absolutely sure that the regions beyond our present horizon are like the parts of the universe we see. As on the ocean, there could be something amazing just over the horizon.

Physicists John Barrow and Frank Tipler have pointed out that a new source of energy—so-called shear-energy—would become available if the universe expanded at different rates in different directions. This shearing of spacetime itself could power the diaphanous electron-positron plasmas forever, if the imbalance in directions persists. To harness it, life (whatever its form) would have to build "engines" that worked on the expansion of the universe itself.

Such ideas imply huge structures the size of galaxies, yet thin and able to stretch, as the spacetime they are immersed in swells faster along one axis than another. This motor would work like a set of elastic bands that stretch and release, as the universal expansion proceeds. Only very ambitious life that has mastered immense scales could thrive. They would seem like gods to us.

As well, our universe could eventually be crushed by denser material not yet in view. Or the smoothing out of mass on large scales might not continue indefinitely. There could be a new range of structures, on scales far larger than the part of the universe we have seen so far. Physics can tell us nothing of these yet. These ideas will probably loom larger as we learn more about the far future of all visible Creation. The accelerating expansion might itself accelerate, leading to the "big rip" that shreds atoms, erasing all information—a truly horrifying prospect, if you think Shakespeare's works should live forever. Surely this is a grand, Wagnerian struggle worthy of life in the far future.

Or, even more fundamentally, maybe time itself is a hominid illusion, not fundamental at all. It might rather be an emergent property of some deeper structure to be revealed. Our human temporal anxiety would then be a passing fashion, not a feature of the universal destiny. This idea may be more sobering than even the cold comfort awaiting us way up ahead, beyond the Year Million.

Chapter 14

After the Stars Are Gone

George Zebrowski

I.

We who have escaped local universes know how and why civilizations turn inward as they are overtaken by regional changes, and either perish or look outward once more.

As the era of star formation ends, heavy matter collects in brown and white dwarfs and neutron stars. Proton decay and particle annihilation provide energy. The black hole aeon of widely differing starlike masses follows, but these evaporate as protons decay, leaving a darkened epoch of scattered photons, neutrinos, electrons, and positrons.

Throughout these aeons, as the locality—that is, the local universe—continues its endless one-way expansion, isolated societies regurgitate their pasts, and retreat into variations of their recorded histories, replacing the darkening reality around them as it thins toward (but never fully reaches) a vacuum state. They accept the small difference between how an external world and one animated by a database (simulating perception of external reality) enters a mind. Creating a structure to support consciousness and to secure storage of information is the essential activity, as it has always been for intelligence, but now at ever decreasing but sustainable levels of energy.

Uncounted societies of overlapping minds accumulate in island redoubts, where preserved data burns with a vital remembrance that does not question whether presents live in the past or pasts in presents and futures. Past is present and future, as surviving minds gather into complex eddies of constantly revised pasts, reliving the data, animating and altering it, and entering it in amnesiac states to dampen spell-breaking doubts.

But these excerpts, reanimated from long states of previous growth, out of which intelligence emerged, are now lost in the re-creations, since no comparison with originals is possible. Without new enrichments, present and future reprise the past. Altered histories themselves become unique relics, and are further changed by a kaleidoscope of perceivers who visit the data and retreat from external reality.

The fate of civilizations in the infinite variants of superspace is a response to loss: first of a planet, then a solar system, a galaxy, and finally a local cosmos.

Localities—for only the background superspace is *the* universe—emerge from an illusory bequest, a legacy that reveals itself as a sudden entropic decrease. Above the misery of self-assembling bioplasm's warring exchanges with itself, stars dance through their main sequence in obedience to the choreography of gravity and its music, thermonuclear fire, while awakening intelligence strives to transcend the chaotic thrall of evolution, the slaughterhouse song of flowing blood whose words pass genetic information forward in a lethal relay race.

Civilizations escape the cradles of life and proliferate throughout the snowflake galaxies, surviving the deaths of suns, peering into the enigmas that seem to stir beyond the horizons of fleeing light, building the knowledge to endure as they draw increasing measures of resources, encountering new life as it emerges from the cruelty of planetary cauldrons into a larger state of inconceivability, out of which is teased every answer except the why and what.

And the blackness always presses in—as the sculpting oceanic field of gravity compresses granular matter into transient stars—and the locality is torn apart, beyond any hope of rebirth through collapse. The darkness has waited from the time before the young galaxies gathered and kindled blindly against the cold, their suns passing their strength first to life and then to intelligence.

Finally, only intelligence shines in the frozen fleeing galaxies. In a universe of cyclical expansion and collapse, mind might have resisted differently; but in a locality of infinitely increasing entropy and expansion, intelligence learns to endure at ever-diminishing rates of energy use, designing and redesigning itself to live at ever subtler scales, maintaining awareness and vitality through ever smaller steps toward an infinitesimally distant floor of unattainable cold.

The darkness will reign forever, but absolute cold is unattainable; and in that is victory, immortality, at ever diminishing expenditures of energy that might never be exhausted.

Candles that would have been nearly invisible to the eyes of early locals brighten the darkness, measuring and remeasuring their output, scaling down to pace the universal decline with a stately approach by halves toward an unreachable oblivion—the ultimate triumph of mind, to live as a warm residue on a flattening curvature of spacetime, inside bright virtual islands where all desires are granted by asking, and the true face of cosmic existence is a medusa to be shunned.

But they made no new memories in those endless aeons, in those countless local universes. All remembrance had long since been laid down and made open to all, in continuous variants that were entered and re-entered without significant change, unnoticed by the deliberately amnesiac. They lived in localities whose accumulations had turned experience into memory, and that back into experiences.

In the continuing variants of an infinite superspace, what happens to societies of mind within localities has been described in myriad ways, but each one confronts the same constraints and a waiting dilemma.

What do you do when it all runs out, when the scenery itself is lost? Intelligences retreat into virtual bubbles of experience where they re-create, out of long-held sentiment, the scenery of their pasts, perhaps their original world and its Moon, alone in a black, starless sky. Persons acquire the appearance of people from their histories, selectively chosen and given a setting. Thus they give life back to the past, and this becomes all of life, the only reality left to mind.

A locality already aeons old, devoid of stars, continues to support intelligent structures inside virtual settings that are, outwardly, large clouds of particles, with an inner awareness that clings to all its pasts, because it needs them to live in.

Deep-time societies of mind, part of our own origins, took of their pasts what they wished, reanimating them as glowing realities. These held the first multiples of minds, already aeons old, vast societies within themselves, arenas of knowledge and aesthetic sense striving for ascendancy, washed up out of time by their old desire for a permanent existence, for a history that would never be lost—into a state of hindsight without a future.

Many convinced themselves that whatever realities had ever existed outside their mental structures were in essence no different from those they were recapturing. They had gathered enough of reality to own it, believing it would never produce more than what they had stored.

But external realms differ. Infinity is what the virtual localities lose, blindly preferring, to the possibility of liberation, a prison of their own design, ignoring the fact that no cosmic period occupies a special place—that even at the highest levels of entropy, emergent possibilities await.

We sent early warnings to these castles of illusion that they would face the decay of heavy particles and a "proton pause," or transition, as minds translate from heavier to lighter substrates, for no self-assembled granularity endures forever, but must eventually return to the substructure. To endure, surviving mentalities reconfigured themselves into lighter regimes. This reconfiguration, which could not be permanent but would provide a bridge, we offered to those who might join us beyond their virtual retreats—or be swallowed by the vacuum.

Some showed no concern about their end, already well aware of the desert of memory in which they lived. Amnesia was necessary to achieve a sense of reality, but each return took its toll on the will to persist. Others saw the pause coming, reconfigured, and again closed their doors.

From our high ground in superspace, we communicated downhill into localities, but such links became more difficult as expansion increased. Enhanced quantum entanglement had been easier in earlier times, but became incoherent in the dark aeons. Communication within localities was intermittently attempted, using photons of immense wavelength just after the proton pause, mostly as signals of survival to other localities.

But the countless localities offered up by a generous superspace were beyond our reach, beyond our capacity to offer admonition; only a finite number could ever be entered and warned. We spoke to the oldest of the lifebearing, well past star formation but just beginning to linger in their pasts, summoning worlds out of the bright mirrors of their minds. Many, but not all, ignored or were unaware of the impending heavy-particle dropout, or of how to deal with it, since their study of outward realities had long been forsaken. Irrationality, in the form of amnesia during virtual re-creations, did not seek outward realities, while the suicidals understood the constraints and chose to ignore the decay that was eating away at their substrate.

We knocked on the vast doors of many strongholds, usually receiving no answer. Memories winked out through huge regions of superspace, even as new universes were born from the vacuum and rushed across time toward the same fate.

But many localities faced the great question: what will there be for us outside our retreats? Have we not always raised civilizations out of given nature? What else is there when all of that is lost?

And we answered that there was the genuine *other* not of our own making. We seek the *outside*, not our own *insides.*

But what is there in that waiting *other*, they ask, which

darkens and dies into oblivion? We can have what we wish within ourselves. We can deal with the danger of the loss of matter. What is it that you offer with this worship of the *other*? We will reconfigure and close our doors—there to raise what is not given but of our own making, they said, fully realizing that only forgetfulness would make the past new.

Their question—What else is there?—we answered by saying that in the superspatial infinity there might be regions unlike any others, realities beyond inward-looking configurations of memory. An infinite superspace would be inexhaustible, and that was what must be sought—not a finite resting place of repetition, but a region where new memories might form, where intelligence might look ever outward, away from these places of endless, lingering, beggarly survival.

We argued with many localities, but the two camps remained deadlocked. They endured heavy particle losses and kept to their ways, and still we looked outward, hoping to glimpse fires beyond the night, while they rekindled the youthful universes from which they had swarmed, failed in, then risen, and would not let die in the glow of their minds.

They cried to us, What difference can you show? Where is this *other* beyond our thoughts that can equal what we have brought out of the past? What is to be done? What is there ever to be done? Tell us, if you can!

We answered that localities spark in the infinity of superspace, that we can never know its full extent, but we can travel as far as we wish, asking why this space exists, why it permits an infinity of finite sparks to light and die.

They mocked our thoughts, saying that our questions would forever flee from us, that our nonlocal existence was a delusion, that we were locals like them with no other ground to stand upon.

We replied that we know something of where we all came from, more about who we are, but would not close the question of what to do and wish to become, or where we might still discover to go.

And they cried back, There is nothing left to learn! There is no outside! Somewhere, something has imagined us, and invented our debate!

Thus, they faced us with a lost argument.

Great tragedies occurred among those who did not make it through the losses of heavy particles that swept localities. Many had readied their engines too late against the siege by cold, refusing to confront the incoming cosmic reality. The loss was vast as the material basis of their lives dissolved into lighter particles, which they might have shaped into a continued survival. But what of it? There is an infinity of sparks in superspace. . . .

The questions remain: What is to be done? What is there to do?

As we continue to look back and to gaze outward beyond the localities, we see intelligences seeking to remake the laws of their own locality, failing to see its embeddedness in a superspace that can never be remade. They still seek to seal themselves up forever in the parsimony of their own locality, ignoring the possibility that in time all particles, even the lightest, will decay and disappear, as their space expands into a near vacuum.

For us, the loss of the genuine *other*, the greater realm that we still understand so poorly and seek to explore, would be catastrophic. Even the revision of natural constraints within individual expansions would not long satisfy finite intelligences, who would find the laws too rigid, would long for a creative regime. Some localities have learned to induce new localities out of their own, but this too promises only a repetition, a local phase-change of physical laws within larger limitations.

Reason, we have learned, is a sharp instrument that cuts infinity into graspable pieces. Large final answers might be incomprehensible or bring catastrophic boredom. Finite beings require infinity around them, with a conflict between defined and undefined moments. Collapse this uncertainty, experienced as time's tension, and we would exist differently, mechanically, losing acute angles of awareness. Reason is a finite angle of perception.

There was a gathering, whose communications we mediated, of some who would revise their localities and those of us who understand the need for a greater mysterium—of those who would live in impoverished virtuals and those who would set out into the unknown. Some defenders of virtuality again made the claim that our vision of infinity was merely another virtual construction imprisoning our thoughts. Thus they granted our warnings but insisted that nothing else was possible.

The infalling ones argued that ours is worship of the unknown and unforeseeable, which amounts to a slavery no better than that of a humiliating existence in given natures, during the aeons when youthful intelligence did not know itself. We replied that new hopes and further growth await us beyond the realm of passing localities, that an uncreated infinity, sufficient to itself, can never be exhausted. They might reshape, even engender new localities, but we will have to face the great mysterium of superspace.

We left them to their fates. The pauses of heavy particles continue, and new localities arise in superspace, each moving toward a discovery of the same alternatives, to stay inside or look beyond. New multifolk emerge from the infinity of locals, and enter our dialogue as the pauses overtake their spaces and their questions compel their thoughts. . . .

But it's always a new landfall on the shores of our infinity, because we can never penetrate very far inland; there is always too much forever ahead. A billion or a trillion epochs, and still the exploration will be infinite. That is the inconceivable reality of a standing infinity—that it is there all at once. You might doubt that it is there, and know that no matter how much of it you cover there can be no certainty that it will go on forever, even as you imagine that it might end in the next few moments. You will never settle the question. And you don't want it settled. An infinite superspace is unique. To discover it as finite would only suggest another beyond—or perhaps a literal nothing.

Infinity is the necessary background for all finite foregrounds; without it, all reasoning leads to infinite regresses of explanation, to inconceivables. Mathematical ways generate these as infinities and have to be renormalized for the minds of finite intelligence. Infinities have to be excluded from the mathematics for it to have use, yet the reality of infinities is inescapable. They are always generated at the extremes of reasoning and cannot help but reveal the character of the larger scheme.

Our lookouts stand watch at the frontiers, hoping to see a dawn across the desert of darkness . . . hoping to glimpse a new, unmemoried *other.* Untried directions await to unfold before us.

II.

What have we envisioned here? Parts of it are rooted, with inevitable failings, in physical measurement, experiment, and observation, and were constructed with considerable intellectual brio by Fred C. Adams and Gregory Laughlin.[1] Since this is a speculative but nonfictional forum, permit me to suspend the narrator's spell and ask:

What are we to make of visions that seem to fall out of the deepest of times into our constrained minds, despite being derived from several ways of physical thinking in today's cosmologies, and combined, as I have done here, with various cultural tendencies and emerging technologies of synthetic experience?

It was much simpler once. Stars died, the universe expanded and collapsed, and did not continue to develop in unforeseeable ways, and the plurality of universes in an infinite superspace was a speculative extravagance—much less, universes that might change their laws through phase shifts.

But our cultures have always lived in virtualities that attract us more than clumsy reality, from which we flee in sleep, and reconfigure as the virtualities of books, plays, poems, movies, games, performances of any kind. We can already see more direct means of sensory input that simulate our incoming real world—with changes. People have long been "moving into" virtual worlds. Go talk to a daytime television soap-opera fan. Some are even earning money out of simple forms of cyberworld building. "Movers" are offering their services. Land values are booming in cyberspace.

In daily reality, by contrast, nothing seems truly to begin or end, nor to have more than a muddle in the middle. Yet story is our basic form of *explanation*, in the sciences and in the arts, a way of revealing why people and things got to be the way they are and what they might become: description and prediction—choose your spectrum of complexity, from things to people. Civilizations have always created virtual bubbles, as they grew more sophisticated in giving people their heart's desire.

The purpose of a civilization is the feathering of nests, as Omar Khayyám confessed long ago:

> Ah, Love! could thou and I with Fate conspire
> To grasp this sorry Scheme of Things entire,
> Would not we shatter it to bits—and then
> Re-mould it nearer to the Heart's Desire!

Who among us wishes to face reality, except to change it? The ones who face reality, as Kurt Vonnegut has said, are those who forget to duck. What Omar Khayyám wrote is exactly what civilization has always been hard at work doing—but there are roadblocks on the road of wishes, because travel on it requires blindness to what is actual and possible. Today's lawyerly forms of discussion seem to allow us to have exactly what we want in a conclusion, if we frontload the argument.

The longing for the assimilation of the *other*, of a person or nature, seems to belong to biology's mill, to the principle of the hybrid organism as a more vital, flexible hold on the future. The same principle sublimates into social life, into the conflict of human groups and civilizations, and perhaps will persist into a collision of intelligent "humanities" throughout the cosmos.

Here I can only glimpse a cosmic drama of familiar things turned unfamiliar. Consider how the constraints of our existence summon for us the observer's spell, the naive reality of spacetime, a sense of direction, space itself, and time that flows but seems not to exist apart from biology and physics. Yet we suspect states of nonlocality, of entanglement with an invisible world that makes all that is visible around us possible at the micro and macro physical scales. As our direct and indirect knowledge deepens and breaks the spell, we long for the answers to the "really good questions," as Rudy Rucker once said to me, the ones our educators pointedly, revealingly, avoided as we grew up, saying, "Go ask your priest or rabbi about that," which so many scientists have been too modest to attempt. But some do attempt them, as John Wheeler and Stephen Hawking have done, and remain modest about the certainty of their answers.

Theories are often not generalizations from observations; as Einstein's "free creations of the mind," they *seek* the observations, because science comes at knowledge from both directions—observations and speculations alike. When you are familiar enough with the rigors of a science, Feynman once said, you have the right to guess. But then, as Einstein agreed, you must still rejoin experience, confirm the speculation with an experiment, a prediction, and repeat it often. We should not ignore a guessing imagination—atomic theory was a 2,400-year-old guess. Niels Bohr once said to a student that his theory was crazy, but not crazy enough to be true.

Maybe there are unique directions in what we call superspace—or, if superspace is too literal a concept, then *regions beyond our eddies of matter* might be a better description. Perhaps our universe is *the* universe, as the logical meaning of the word would have it, with nothing else beyond it, truly *nothing*—as much an affront to our reason as infinity, or the impossible existence of a genuine nothing. How can an infinity be unique and still be called an infinity? And yet it seems that perhaps there is only one, in our way of talking, and there can be only one infinity. Medieval scholars warned of madness in such thoughts and imagined that *all* is only a small bean in the palm of one's hand. Cantor's theory of transfinites would have seemed the devil's work.

Here I ask you back into the cosmic scenario, which sits up there in deep time, telling itself to its past.

Structure and communications, so much easier today, become more difficult in the darkest deeps of time, which, in our present ignorance, become sketchy. "The seeming poverty of this distant epoch," according to Adams and Laughlin, "is perhaps due to the difficulties inherent in extrapolating far enough into the future, rather than an actual dearth of physical processes."[2] Much more can go on, they argue, which we just can't imagine today. Sounds hopeful.

In my narrative, the constrictions seem greater because realities are assumed. But these difficulties, even if they can be imagined away with some justification, amount to profound ignorance. *Physicality*, the "stuff" of things, becomes vague and strange.

My narrator, and secret self, navigates superspace and communicates with localities, perhaps even enters them physically. Faster-than-light versions of various particles, translations of structure from one kind to another, hover in the background as enablers of possibility, as do vast energies tapped from the superspatial vacuum. Curious that so much can be imagined, with some basis in doable deeds, even if it does run us off our maps of ignorance. Einstein thought in curious pictures, then in fictional "thought experiments," then got some math and a Laputan raft of scientists to do the experiments. That we can work back from some imaginings and thought experiments to mathematics and physical experiment might be the most important thing we do as a species. Before anything can be done, it must first be imagined, however imperfectly. In a sea of photons, the waves can be as long as a universe.

So what happened to all the cultures as the scenery faded away? Those that did not perish in much earlier times turned inward, adjusted their use of available energy, their eyes to darker hues, and lived forever, as Freeman Dyson has told us, and had time to ask all questions and find the answers in the endless twilight of the red dwarfs, according to Arthur C. Clarke.[3]

Some kept a window to the outer darkness and looked out from their brightly colored inner realities into the abyss, and only vaguely realized that there was anything to be done about it.

Groups gathered from time to time in their redoubt, as if for an all-night dinner party, and talked for what might have been a million years, a short time by other measures, as the black infinity stood waiting in their windows, reproaching them.

I hear their dinner conversation, in whatever way it is that they speak, as they recount narratives from the star-bearing age of our universe. They might as well be a group of friends playing cards in a basement room, with a warm August rain running on the small windows, where it hasn't rained in trillions of years, but the rain remembers.

III.

The stories have one thing in common—that theirs is a central time, even after the stars were gone and all knowledge and all narratives were gathered for endless re-examination and re-experience in endless variations, until the will to look outward re-emerged, when the romance of reworking pasts was over.

That these epochs were as they were, starry and then dark, is not strange to us who see that a richly developing outwardness accumulated its knowledge and experience and then became a rich inwardness even as the universe darkened and thinned, until that inner refuge was in turn used up, forcing the search for a new dawn, even a remaking of local natures, as each alternative unveiled itself only when the previous one had been exhausted by intelligent life's hungers. Digestion of the past became a way forward.

Anxious apprehensions spoke as if from the future to the past, drawing the past forward, lest it hurtle blindly to some final catastrophe.

The midnight dinner party was a relic of ancient pasts, where the guests mirrored and overlapped each other and sought revelations in each others' deeps.

One day they heard a great knocking on the door to their virtual realm. Our messenger delivered the warning about the heavy-particle pause, and it was heeded. After the translation into a lighter state, the debate continued over whether to remain inside or surge outward.

Our twilight, said one of the debaters, was prefigured in epochs past, in the short-lived creatures, who when they lay dying reclaimed their happiest memories, substituting internally stored data for external inputs (for when output is connected to input, omnipotence is experienced). With sufficient richness gleaned from experience and with a mask of amnesia added, there was no critical difference between fantasy and reality. A curious fact, that omnipotence might be simulated. . . .

An old daughter hurried her ancient mother to bed and heard the mother's happy times with her dead husband replaying in her dreams. Intelligence takes experience and turns it into memoried information. "We're making memories," people say, "laying them down for later." In ourselves, and stored elsewhere. The word of beginnings becomes the word again. . . .

But those who still looked outward, even when faced with a dwindling input from a darkening universe, stubbornly preferred the given to the remembered. Outwardness deserved greater respect and perusal and would, they claimed, yield rewards larger than any inner recall. The past was a reef of dead accumulations, and despite its beauty would not provide a novel futurity.

Out there, purposeless, enigmatic, the greater realm stood away from us forever . . . gifting the survivors of local realms with its provocation.

Still, the tendency of cultures to make interiors persisted, even as the outside kept breaking in, mocking the sought-for independence of mind. Might not universes be created, needing no greater background to exist—to be unaffected by any constraints outside their boundaries?

How often came that terrible knocking on the great virtual doors by something outside, and those inside knew the importance of their door and a meaningless fear of the pounding? It was an old knowledge, without specific meaning, but suggesting more than all the old knowledge and experience hoarded inside. The thing outside. Not because it was any specific thing, but because it was ***outside***.

To know from the outside—of a culture, of a local universe, but not from outside of infinity—was to know endless fear, to be both free and lost, to stand at the edge of the abyss, to teeter and not fall, to fall and hold back, to be both damned and saved. But damned for what? Saved for what? Fearful of one or the other, or both.

IV.

There is not enough inner life in our speculations about futures—near, middle distance, or deep timed. Our cortex is a shallow, prepubescent child that dreams of reason's independent, militant, even repressive way of seeing, as nightmares and predatory survivalist scenarios swirl beneath its feet, and reason that stands aside and confesses this to the reader with the words you are reading is a traitor to a rationalizing mixture of intellect and impulse. But there is hope in that betrayal of our natures. To paraphrase C. M. Kornbluth, our speculations about even the deepest of futures are weighed down by the individual's relationship to his family, by instinctive feelings, and by superstitious intuitions about the raw universe around us.[4] We long to be proven wrong about its gross "materiality," its lack of "noble spirit," that in all its grandeur it might be nothing more than a "great thought," as the cosmologist Sir James Jeans famously put it,[5] or perhaps a greasy spot on a wall near the floor of some larger existence, whose intelligent life has the same problem of explaining itself as we do.

We return, in this realm of words where evocative logicians know how to work a conclusion, to a possible cosmic model of our universe, the one and only universe, which expands without end and resists thinning into vacuum, but continues to evolve in strange, creative ways, according to chaos theory's wondrous ways. The most interesting point in Adams's and Laughlin's original paper comes at the end, when they announce the "Principle of Copernican Time," which holds that "the current cosmological epoch has no special place in time . . . interesting things can continue to happen at the increasingly low levels of energy and entropy available in the universe of the future."[6] If conscious structure and information can be conserved, the implications for the *insides of mind* offer an endless hope. We seem to need the endless at every morning, and need not fear it as Pascal did his intruding infinity.

Beyond this hope still wait the inarticulate dragons of *why.* They live in that great emptiness between premises and conclusions, which we do not know how to fill when it comes to the inconceivabilities, in order to avoid tautologies, truisms, insistence and begged questions, dogmas and matters of faith—all of which are one and the same stubborn rhetoric of instinctive life.

Intelligent life *knows* that it exists *inside* something, and suspects why it cannot see or know what it most desires—*final answers*, which might in fact belong to a mistaken desire or a wrong way of asking. A right way would require a metaphysical prison break of transcendent outcome, modeled, perhaps, on the breakouts we have made from our own previous conceptual *insides*—the escapes made by Copernicus, Galileo, Darwin, Gödel, Einstein, and E. O. Wilson. All visited humiliation on those unwilling to break with previous universes—which seem to change with our own changing insides.

Inside, and even deeper inside the redoubts we have built against nature, we not only suspect that we have been asking the wrong questions about a fictional outside, but that there are no questions to ask beyond the sightedness of our angular perspectives, which seem essential to our intellects.

As I look at my shelves of fiction and science fiction, at the DVD files of movie dramas and CDs of strangely articulate audio compositions that sound like prayers, I see virtual realities in training. We *are* moving into a reality of our own making and, however deficient it might be, we will close the doors—and perhaps continue to argue whether physical infinities are of greater interest than these prosthetic contrivances of our own making that we call civilizations. We question infinity but flee it, even as we exhaust our physical environs and inner ingenuities.

Mileva Marić, Einstein's first wife and a physicist, wrote to her future husband: "I do not believe that the structure of the human brain is to be blamed for the fact that man cannot grasp infinity. Man is very capable of imagining infinite happiness, and he should be able to grasp the infinity of space."[7]

Perhaps we are not a breakout species. But as Kurt Gödel might have said, there is hope in our ability to make such a statement, as we live with ignorance that is itself a map.

These words, composed of fictions that almost certainly cannot match a complex reality (that thing, as Philip K. Dick once said, that persists after you stop believing in it), have been knocking not at a door but at a wall around human minds. We know it's there, but it seems important to step up to it, however inarticulately, to know that at least we have come to it. Once, we simply made it all up, with stories in place of physical observations and revealing experiments.

But we who live Gödel's forever-incomplete reality might learn a lot and never get far into infinity, go as far in as we wish and not get far, knowing that it started a long way behind us. Would you wish to exist in an exhaustible universe? Well, perhaps for a moment or two, just to know what the ultimate answers are like—and then to forget them.

We have exhausted twenty or thirty civilizations, made and lost, every one a failed prosthesis fitted onto nature's otherness.

Shall we remake nature or turn our backs on it? Can any culture make of itself something rich and inexhaustible? Then it will have made a cosmos—except that there cannot be more than one such, it seems. The pride of Lucifer knew no bounds, and perhaps that is what he made for himself, to reign in his own hell rather than serve heaven's givens.

Superspace, as we have understood its infinity with our words, would be *the only* universe—uncreated and inexhaustible, requiring no further explanation. That's the same pass we give to any conception of God. The buck of causes and explanations stops there. Perhaps it does not need to stop; *we* need stops, renormalizations of infinity, according to profane reason, which tailors reality as best it can. Is there another kind?

When we have explained the particles as being such and such, we would have to explain what *that* is, and we're off on an infinite regress, with each explanation leading to a new question. Common language also yields infinities—as does mathematics—and infinities must be dealt with. Either the regress is the final answer, or we are asking the wrong question and must ignore the infinity, accepting only answers that do not produce infinities. This seems a frustrating impasse, an insult to both intuition and reason. In mathematics, the elimination of infinities is seen as peculiar, odd, even an unfair ad hoc procedure, which nevertheless yields useful results. This suggests both what kind of universe we live in *and* the nature of our minds. The usefulness of "renormalization" is not curious when we recall the scalpel-like selectivity of our minds, in the tools of distinction and definition so clearly seen in our logical and mathematical notation. To hit an infinity in mathematics, or an infinite regress in a common language argument, discredits the effort's usefulness. There's the wall, so to speak, around the kind of open universe we live in. We know it and we won't.

Medieval mystics saw the universe as a bean held in a cosmic palm, perhaps ready for a soup. Faced with a clearly defined, final answer to the nature of the universe, perhaps we would not rejoice but instead go mad—because then, finally, it could *never* be anything else.[8]

Notes

Introduction

1. Peter Ward, ***Future Evolution: An Illuminated History of Life to Come*** (New York: W. H. Freeman, Times Books, Henry Holt and Company, 2001), 167.
2. Dougal Dixon, ***After Man: A Zoology of the Future*** (New York: St. Martin's Press, 1981/1998), 32.

Chapter 1: "The Laughter of Copernicus" by Jim Holt

1. J. Richard Gott, "Implications of the Copernican Principle for Our Future Prospects," ***Nature,*** May 27, 1993.
2. J. Richard Gott, ***Time Travel in Einstein's Universe*: *The Physical Possibilities of Travel through Time*** (Boston: Houghton Mifflin, 2000), 291.
3. Robert R. Provine, ***Laughter: A Scientific Investigation*** (New York: Viking, 2000), 75 ff.
4. Richard Dawkins, ***The Ancestor's Tale: A Pilgrimage to the Dawn of Evolution*** (Boston: Houghton Mifflin, 2004), 100.
5. Stanislaus Dehaene, ***The Number Sense: How the Mind Creates Mathematics*** (New York: Oxford University Press, 1997), Chapter 1.
6. Bertrand Russell, ***The Autobiography of Bertrand Russell*** (London: Allen & Unwin, vol. I, 1967), 45.

7. Bertrand Russell, *Mysticism and Logic* (New York: Doubleday Anchor, 1957), 57.

8. Bertrand Russell, *My Philosophical Development* (New York: Simon and Schuster, 1959), 157.

9. Don Zagier, "The first 50 million prime numbers," *Mathematical Intelligencer*, (1977): 8.

10. Marcus du Sautoy, *The Music of the Primes: Searching to Solve the Greatest Mystery in Mathematics* (New York: Harper Perennial, 2003), 114.

11. Karl Sabbagh, *The Riemann Hypothesis: The Greatest Unsolved Problem in Mathematics* (New York: Farrar, Straus and Giroux, 2002), 42.

12. John Derbyshire, *Prime Obsession: Bernhard Riemann and the Greatest Unsolved Problem in Mathematics* (New York: Plume, 2003), 54.

13. Marcus du Sautoy, *The Music of the Primes: Searching to Solve the Greatest Mystery in Mathematics* (New York: Harper Perennial, 2003), 167.

14. Karl Sabbagh, *The Riemann Hypothesis: The Greatest Unsolved Problem in Mathematics* (New York: Farrar, Straus and Giroux, 2002), 135.

15. G. H. Hardy, *A Mathematician's Apology* (Cambridge: Cambridge University Press, 1940), 150.

16. Alain Connes and Jean-Pierre Changeux, *Conversations on Mind, Matter, and Mathematics* (Princeton, N.J.: Princeton University Press, 1989), 112.

17. Bertrand Russell, *Nightmares of Eminent Persons, and Other Stories* (Harmondsworth: Penguin Books, 1955), 46.

18. V.S. Ramachandran, with Sandra Blakeslee, *Phantoms in the Brain: Probing the Mysteries of the Human Mind* (New York: Harper Perennial, 1998).

Chapter 3: "A Million Years of Evolution" by Steven B. Harris

1. The prototype fossil for very early humans already nearly modern in their postcranial skeletons is the so-called Turkana Boy (KNM-WT 15000), a teenaged male who died about 1.5 million years ago, and whose skeleton was found almost completely intact.
2. Frank W. Marlowe, "Hunting and Gathering: The Human Sexual Division of Foraging Labor," *Cross-Cultural Research*, 41, 2 (2007): 170-195. DOI: 10.1177/1069397106297529. SAGE Publications.
3. This assumes a neuron with a typical volume of five thousand cubic microns. The size of neurons is quite variable, however.
4. Such creatures are not at all what Norbert Weiner had in mind when he coined the word *cybernetic* (see *The Human Use of Human Beings: Cybernetics and Society*, Boston, Houghton Mifflin, 1954), but we're stuck with the alternate meaning of the word as referring to a human-machine meld, and there's little we can do about it.
5. See http://fusedweb.pppl.gov/CPEP/Chart_Pages/5.Plasmas/SunLayers.html. Accessed October 25, 2007.
6. 12,000 solar-output equivalent years = 0.0012 x (1/8) x [80 billion years/1000]. The 0.0012 is the mass fraction of deuterium in Jupiter's hydrogen, the one-eighth factor is due to deuterium fusion being only about one-eighth as efficient as light hydrogen fusion, on a per-mass basis; 80 billion years is the theoretical time to burn all of the Sun's hydrogen; and the factor of 1,000 corrects for Jupiter's smaller mass.
7. The rate at which power is returned from deuterium mining *P* as a function of the time t that a shell of mining probes moves outward and returns deuterium at speed *v*

(assumed the same return speed as the probes move), into a galaxy with stars at a volumetric density ρ, where each star with a mean deuterium energy supply A, is given by: $P = \pi \rho A v^3 t^2$. If **A** is given in solar-output-years of energy, ρ in stars per cubic light-year, v as a fraction of c, the speed of light, and t is in total years of probe outward exploration, then power P will be given in units of solar-output years of energy returned, per year, at time t in the future. Removing A from the equation and multiplying by a factor of 4 (to remove the effect of energy-return time, which no longer cuts the rate) gives the number of stars explored per year by the probe shell. The figure given in the chapter assumes a star density of 0.00024 stars per cubic light year (this is hurt by the flattened distribution of stars in the galactic plane), a probe and deuterium return speed of 0.01 c, and an average of 1,000 solar-output years of deuterium in the gas giants of each target star (as compared with an estimate of 12,000 solar-output years contained in Jupiter).

8. For slower speeds, there is a bit of help in that aerocapture is a gravity-well maneuver, so that speed loss needed in a photosphere is not full interstellar speed, but only the speed associated with energy in excess of solar escape energy at photosphere distance. For example, a probe crossing between Sun-like systems at 0.002 c will arrive at the target photosphere at 0.00283 c (not .004 c), and thus need shed only 0.00083 c for capture, which is less than 42 percent of its interstellar speed.

9. Or (Sagan was an optimist), perhaps some aliens learn to get along, and these are the ones we'll hear. When we tune into what they're saying—sometimes Sagan's optimism could be wild—their communications might contain helpful tips on how avoid nuclear annihila-

tion [see Carl Sagan, "The Quest for Extraterrestrial Intelligence," *Broca's Brain* (New York: Ballantine Books, 1979), Chapter 22].

10. Computer power is expressed in FLOPS (FLoating Point Operations Per Second). The largest computers today are in the petaflop range (10^{15} operations per second). A yottaflop is a billion times faster, and is the fastest operational power for which we currently have a defined word.

Chapter 4: "Life among the Stars" by Lisa Kaltenegger

1. S. Udry, X. Bonfils, X. Delfosse, T. Forveille, M. Mayor, C. Perrier, F. Bouchy, C. Lovis, F. Pepe, D. Queloz, and J.-L. Bertaux, "The HARPS search for southern extra-solar planets. XI. Super-Earths," *A&A* 469 (2007): L43–L47

2. L. Boisnard and M. Auvergne, "The CoRoT Mission—Pre-Launch Status," *ESA SP* (2007): 1306.

3. W. J. Borucki, Koch, D. G., Basri, G. B., Caldwell, D. A., Caldwell, J. F., Cochran, W. D., Devore, E., Dunham, E. W., Geary, J. C., Gilliland, R. L., Gould, A., Jenkins, J. M., Kondo, Y., Latham, D. W., Lissauer, J. J., "Scientific Frontiers in Research on Extrasolar Planets," ASP Conference Series, Vol. 294, ed. Drake Deming and Sara Seager (San Francisco: ASP, 2003): 427–440.

4. M. Fridlund, "DARWIN: The InfraRed Space Interferometer," *ESA-SCI* (2000): 12, 47.

5. C. A. Beichman, N. J. Woolf, and C. A. Lindensmith, "The Terrestrial Planet Finder (TPF): A NASA Origins Program to Search for Habitable Planets/the TPF Science Working Group," *JPL publication* (Washington, D.C.: NASA JPL, 1999): 99–103.

6. H. Levine, S. Shaklan, and J. Kasting, "Terrestrial Planet Finder Coronagraph Science and Technology Definition Team (STDT) Report" (Pasadena, Calif: Jet Propulsion Laboratory, 2006).
7. C. Sagan, *COSMOS* (New York: Random House, 1980).
8. L. Kaltenegger, W. A. Traub, and K. W. Jucks, "Spectral Evolution of an Earth-like Planet," *ApJ* 658 (2007): 598–616.
9. J. F. Kasting, D. P. Whitmire, and R. T. Reynolds, "Habitable Zones around Main Sequence Stars," *Icarus* 101 (1993): 108–128.
10. F. Selsis, J. F. Kasting, J. Paillet, B. Levrard, and X. Delfosse, "Habitable planets around the star Gliese 581?," *A&A* (2007): 1373-1387.
11. L. Kaltenegger, A. Segura, J. Kasting, W. Traub, K. Jucks, *ApJ* (in press, 2008).
12. A. P. Boss, "ASP Conference Proceedings," *Astronomical Society of the Pacific*, 219, 7; ed. F. Garzón, C. Eiroa, D. de Winter, and T. J. Mahoney.
13. A. Leger, F. Selsis, C. Sotin, T. Guillot, D. Despois, D. Mawet, M. Ollivier, A. Labeque, C. Valette, F. Brachet, B. Chazelas, and H. Lammer, "Ocean-Planets," *Icarus* 169 (2004): 499–504.
14. L. Kaltenegger, W. A. Traub, and K. W. Jucks, "Spectral Evolution of an Earth-like Planet," *ApJ* 658 (2007): 598–616.
15. F. Selsis, J. F. Kasting, J. Paillet, B. Levrard, and X. Delfosse, "Habitable planets around the star Gliese 581?," *A&A* (2007): 1373-1387.

Chapter 5: "A Luminous Future" by Catherine Asaro

1. Letters to the Editor, *Scientific American*, February 1996, 10.
2. Neil DeGrasse Tyson, "The Beginning of Science," *Natural History* (March 2001) or www.research.amnh.org/~tyson/18magazines_beginning.php.
3. Wolfram Research, scienceworld.wolfram.com/biography/Kelvin.html.
4. C. Asaro, "Complex Speeds and Special Relativity," *American Journal of Physics* (April 1996).
5. O. M. P. Bilaniuk, V. K. Deshpande, and E. C. G. Sudarshan, "'Meta' Relativity," *American Journal of Physics* 30 (1962): 718–723.
6. G. Feinberg, "Possibility of Faster-than-Light Particles," *Physical Review* 159 (1967): 1089–1105.
7. O. M. Bilaniuk and E. C. G. Sudarshan, "Particles Beyond the Light Barrier," *Physics Today* (May 1969): 43–51.
8. M. N. Kreisler, "Are There Faster-than-Light Particles?" *American Scientist* 61 (1973): 201–208.
9. A very readable review of early experimental work appears in G. Feinberg, "Particles That Go Faster than Light," *Scientific American* 222 (February 1970), 68–77. An intriguing discussion of what happens to the positions of faster-than-light particles is given by E. Recami and R. Mignani, "Generalized Lorentz Transformations in Four Dimensions and Superluminal Objects, " *Nuovo Cimento* 14 A (1973): 169-189. Relativity buffs interested in more on treatments of the subject using general relativity might enjoy the science fact article "Faster-than-Light" by Dr. Robert Forward in the February 1994 issue of *Analog* magazine.

10. See the columns by John Cramer and references therein: www.npl.washington.edu/AV/altvw81.html,

    www.npl.washington.edu/AV/altvw86.html, www.npl.washington.edu/AV/altvw99.html, www.npl.washington.edu/AV/altvw33.html, www.npl.washington.edu/AV/altvw39.html, www.npl.washington.edu/AV/altvw103.html, www.npl.washington.edu/AV/altvw62.html,
11. The contraction actually appears as a rotation for parts of an object that subtends an infinitesimal solid angle as seen by the observer. This rotation is an approximation for infinitesimal angles, as can been seen by imagining a relativistic train traveling along rails which are at rest relative to the observer. The train cannot rotate while the rails remain fixed. See J. Terrell, "Invisibility of the Lorentz Contraction," *Physical Review* 116 (1959): 1041–1045.
12. When we talk about an observer recording these strange effects, it raises the question of *how* she makes measurements when we are going so fast. We will assume she can take data infinitely fast. Physics books often show helpful pictures of an observer comfortably seated in a chair while a rocket or relativistic train obligingly passes by in front of the chair, somehow being clearly visible despite its phenomenal speed.
13. The equations of special relativity can be found in any physics text with an introduction to modern physics. See for example R. A. Serway (*Modern Physics*, New York: Thomson Brooks/Cole; Third Rev Ed, 2005). For a detailed online discussion of relativity and its applications to the ideas of science fiction, see http://www.physicsguy.com/ftl/html/FTL_intro.html.

14. Historically, m° has been called the rest mass. Bilaniuk and Sudarshan point out that the word is somewhat of a misnomer for tachyons given that a superluminal particle has no rest frame (see note 7 above). They suggest the term "proper mass."

15. Making speed complex nicely solves a problem that has long plagued science fiction writers, which is how to design faster-than-light spaceships. The idea inspired my story "Light and Shadow" in *Analog, Science Fiction and Science Fact*, April 1994 and my novel *Primary Inversion* (New York: Tor, 1995).

Chapter 6: "Citizens of the Galaxy" by Wil McCarthy

Author's note: These concepts have been written about extensively by a number of science fiction writers, with the best expressions coming from Stephen Baxter, David Brin, Greg Egan, David Gerrold, Linda Nagata, Bruce Sterling, Charles Stross, and John C. Wright. I will immodestly include my own name as well.

1. This heroic enterprise is sketched in more detail at www.scifi.com/sfw/issue423/labnotes.html and in Wil McCarthy, *To Crush the Moon* (New York: Bantam Books, 2005), Appendix C: Technical Notes.

2. Wil McCarthy, *Hacking Matter* (New York: Basic Books, 2004). This is available as a free download at www.wilmccarthy.com/hm.htm.

Chapter 7: "Under Construction" by Robert Bradbury

1. This suggests a sophisticated extra factor needed in the Drake equation commonly used in SETI to evaluate the probability and abundance of alien life.

2. There were other alternative paths, paths that included a more subtle harvesting of solar energy, paths which

would not allow external observers to judge the state of one's stellar civilization. Some might have followed such paths, and must therefore be incorporated into simulations of the state of the visible universe.

3. The assumption here is that the sort of fax-like teleportation described in the previous chapter proves to be impossible.

References

R. J. Bradbury, "Matrioshka Brains" (1997). http://www.aeiveos.com:8080/~bradbury/MatrioshkaBrains/MatrioshkaBrainsPaper.html

D. R. Criswell, "Solar System Industrialization: Implications for Interstellar Migrations," in *Interstellar Migration and the Human Experience*, eds. Ben R. Finney and Eric M. Jones (Berkeley, CA: University of California Press, 1985): 50–87.

K. E. Drexler, *Nanosystems: Molecular Machinery, Manufacturing and Computation* (John Wiley & Sons, 1992).

———"Molecular Manufacturing for Space Systems: An Overview," *J. British Interplanetary Society*, 45 (1992):401–405.

F. J. Dyson, *Disturbing the Universe* (New York: Harper and Row, 1979).

———"Search for Artificial Stellar Sources of Infrared Radiation," *Science* 131, 3414 (June 3, 1960): 1667–1668.

H. Moravec, *Mind Children: The Future of Robot and Human Intelligence* (Cambridge, MA: Harvard University Press, 1987).

K. G. Suffern, "Some Thoughts on Dyson Spheres," *Proc. Astronomical Society of Australia*, 3(2):177-179 (1977).

Chapter 8: "The Rapacious Hardscrapple Frontier" by Robin Hanson

1. A. Paultre, Definition of "Hardscrapple," *Everything 2,* http://everything2.com/index.pl?node_id=1509110, December 28, 2003.

References

Hanson, "Burning the Cosmic Commons," working paper, http://hanson.gmu.edu/filluniv.pdf, July 1, 1998.

"Long-Term Growth as a Sequence of Exponential Modes," working paper, http://hanson.gmu.edu/long-grow.pdf, December 2000.

Chapter 9: "Do You Want to Live Forever?" by Pamela Sargent and Anne Corwin

1. R. Pells, "History Descending a Staircase: American Historians and American Culture," *Chronicle of Higher Education*, 53, 48 (Summer 2007).
2. http://www.legacy98.org/timeline.html, "In Griswold v Connecticut, 381 U.S. 479 (1965), the Supreme Court overturns one of the last state laws prohibiting the prescription or use of contraceptives by married couples."
3. Aubrey de Grey, Ph.D., *Ending Aging* (New York: St. Martin's Press, 2007), 42–43.
4. http://methuselahfoundation.org/index.php?pagename=lysosens.
5. http://methuselahfoundation.org/index.php?pagename=mitosens.
6. http://www.foresight.org/Nanomedicine/SayAh/index.html.
7. http://mr.caltech.edu/media/Press_Releases/PR12745.html.

8. http://www.sciencedaily.com/releases/2007/03/070308220454.htm.

9. R. Tallis, "Enhancing Humanity," *Philosophy Now*, 62, (July/August 2007).

10. F. Dyson, "Our Biotech Future," *The New York Review of Books*, 54, 12, July 19, 2007. Dougal Dixon's inventive volume *After Man: A Zoology of the Future* (New York: St. Martin's Press, 1981), might provide a useful guide for such projects.

Chapter 10: "Communicating with the Universe" by Amara D. Angelica

1. In the late 1990s, Dr. Vint Cerf, who co-designed the Internet's TCP/IP protocol, designed the Interplanetary Internet (IPN, http://www.ipnsig.org) to link up the Earth with other planets and spaceships in transit over millions of miles. Cerf's clever scheme solved a big problem. With interplanetary communication delays—the average two-way latency (delay time) between Earth and Mars, 228 million km apart, is 25 minutes 21 seconds—the Internet TCP/IP protocol we use today would simply time out. Who has half an hour to wait for a carriage return? So Cerf and his team came up with a store-and-forward architecture—a sort of relay race. Transmit messages to an Earth-orbiting satellite, let's say, and store them there until the next local pass of the Moon, which then transmits them to Mars.

2. T. Krag and S. Büettrich, "Wireless Mesh Networking," *O'Reilly Network*, Jan. 22, 2004: http://www.oreillynet.com/pub/a/wireless/2004/01/22/wirelessmesh.html.

3. Energy Information Administration , U.S. Department of Energy: http://www.eia.doe.gov/emeu/international/

electricityconsumption.html. The world electrical power generation is increasing by 2.4 percent per year (see http://www.eia.doe.gov/oiaf/ieo/electricity.html) and is expected to grow to thirty terawatt-hours in the year 2030.

4. Martin I. Hoffert, K. Caldeira, G. Benford, D. R. Criswell, C. Green, H. Herzog, A. K. Jain, H. S. Kheshgi, K. S. Lackner, J. S. Lewis, H. D. Lightfoot, W. Manheimer, J. C. Mankins, M. E. Mauel, L. J. Perkins, M. E. Schlesinger, T. Volk, T. M. Wigley, "Advanced Technology Paths to Global Climate Stability: Energy for a Greenhouse Planet," *Science*, 298 (2002): 981–987. As Dougal Dixon notes in Chapter 2, we are running out of oil, and what is worse, many countries, especially China, are burning huge amounts of coal, increasingly polluting the atmosphere with toxins and carbon dioxide and accelerating global warming. It can only get worse: 850 new coal-fired power plants are to be built by 2012 by the United States, China, and India. Terrestrial solar installations, biofuel, wind power, and geothermal power will help, but they all have limitations (ground-based solar panels don't work at night, for example) and, says Hoffert, can't economically provide the amount of power needed, especially in Africa and Asia.
5. M. Hoffert, "Energy from Space," Marshall Institute, Aug. 7, 2007, http://www.marshall.org/article.php?id=550.
6. John G. Cramer, "Wormholes and Time Machines," *Analog Science Fiction and Fact*, June 1989; John G. Cramer, "EPR Communication: Signals from the Future?," *Analog Science Fiction and Fact,* (December 2006), http://www.analogsf.com/0612/altview.shtml; Max Tegmark, "Parallel Universes," *Scientific American*, May 2003.

7. Seth Lloyd, *Programming the Universe: A Quantum Computer Scientist Takes On the Cosmos* (New York: Knopf, 2006), 165.

8. J. G. Cramer, "An Experimental Test of Signaling Using Quantum Nonlocality," http://faculty.washington.edu/jcramer/NLS/NL_signal.htm.

9. J. G. Cramer, "Reverse Causation and the Transactional Interpretation of Quantum Mechanics, in Frontiers of Time: Retrocausation—Experiment and Theory," in AIP Conference Proceedings, vol. 263, ed. Daniel P. Sheehan (Melville, N.Y.: AIP, 2006), 20–26; J. G. Cramer, "Reverse Causation—EPR Communication: Signals from the Future?" *Analog Science Fiction and Fact* (December 2006), http://www.analogsf.com/0612/altview.shtml; B. Dopfer, Ph.D. thesis, University of Innsbruck (1998); A. Zeilinger, *Rev. Mod. Physics*, 71 (1999): S288–S297.

10. Jack Sarfatti, *Super Cosmos* (Bloomington, I.N.: Author House, 2006), 20.

11. R. A. Freitas, Jr., and F. Valdes, "The Search for Extraterrestrial Artifacts (SETA)," *Acta Astronautica*, 12 (1985): 1027–1034.

12. P. Liljeroth, J. Repp, and G. Meyer, "Current-Induced Hydrogen Tautomerization and Conductance Switching of Naphthalocyanine Molecules," *Science* (August 2003), 317, 5842, 1203–1206, http://www.sciencemag.org/cgi/content/abstract/317/5842/1203

13. C. Rose and G. Wright, "Inscribed Matter as an Energy Efficient Means of Communication with an Extraterrestrial Civilization," *Nature*, 431, September 2004, http://www.winlab.rutgers.edu/~crose/papers/nature.pdf.

14. S. Shostak, "What Do You Say to an Extraterrestrial?" *SETI Institute News*, December 2, 2004, http://www.seti.org/news/features/what-do-you-say-to-et.php.
15. William E. Burrows, *The Survival Imperative: Using Space to Protect Earth* (New York: Forge, 2006).
16. Personal communication, September 3, 2007.
17. Stephen Wolfram, *A New Kind of Science* (Champaign, I.L.: Wolfram Media, 2002), 1188, http://www.wolframscience.com/nksonline/page-1188b-text.
18. M. Chown, "Looking for Alien Intelligence in the Computational Universe," *New Scientist*, November 26, 2005, http://www.newscientist.com/channel/fundamentals/mg18825271.600.
19. H. Muir, "Did Life Begin on Comets?" NewScientist.com news service, http://space.newscientist.com/channel/astronomy/astrobiology/dn12506, August 17, 2007.
20. M. Peplow, "ET Write Home," *Nature News*, http://www.nature.com/news/2004/040830/full/040830-4.html, September 1, 2004.
21. P. Davies, "Do We Have to Spell It Out?" *New Scientist*, August 7, 2004, http://www.newscientist.com/article/mg18324595.300.
22. Ray Kurzweil, *The Singularity Is Near* (New York: Viking, 2005).
23. Fred Hoyle and John Elliot, *A For Andromeda* (New York: Harper & Row, 1962), adapted from the 1961 BBC TV serial, now lost: http://www.imdb.com/title/tt0054511/. 24.
24. James Gardner, *The Intelligent Universe* (Franklin Lakess, N.I.: New Page Books, 2007).

25. In Chapter 7 of this book, Wil McCarthy estimates that people could store most of their memories in about two terabytes, which could be transmitted via satellite in just a few hours.

26. CyBeRev, Terasem Movement, Inc., http://www.cyberev.org.

27. http://en.wikipedia.org/wiki/Computronium.

28. Seth Lloyd, *Programming the Universe: A Quantum Computer Scientist Takes On the Cosmos* (New York: Knopf, 2006), 165.

29. Based on the Margolis-Levitin theorem: take the amount of energy within the horizon (10^{71} joules), multiply by 4, and divide by Planck's constant. What has the universe computed? Itself. S. Lloyd, *Programming the Universe: A Quantum Computer Scientist Takes On the Cosmos*: 165-167.

30. M. McKee, "Black Holes: The Ultimate Quantum Computers?" *NewScientist.com news service*, March 13, 2006, http://space.newscientist.com/article.ns?id=dn8836.

31. "Chandra Finds Evidence for Swarm of Black Holes near the Galactic Center," January 12, 2005, http://www.sciencedaily.com/releases/2005/01/050111114024.htm

Chapter 11: "The Great Awakening" by Rudy Rucker

1. Seth Lloyd, *Programming the Universe: A Quantum Computer Scientist Takes On the Cosmos* (New York: Knopf, 2006).

2. For details on this point, see Rudy Rucker, *The Lifebox, the Seashell, and the Soul* (New York: Thunder's Mouth Press, 2005), or see the topic "irreducibility" in Stephen

Wolfram, *A New Kind of Science* (Champaign, I.L.: Wolfram Media, 2002).

3. See David Skrbina, *Panpsychism in the West* (Cambridge: MIT Press 2005).
4. See Antonio Damasio, *The Feeling of What Happens* (New York: Harcourt, 1999), and Jeff Hawkins and Sandra Blakeslee, *On Intelligence* (New York: Times Books, 2004).
5. Robert Sheckley, "Specialist," from his landmark collection, *Untouched by Human Hands* (New York: Ballantine Books, 1954).

Chapter 13: "The Final Dark" by Gregory Benford

1. The literature of science fiction has always looked long, from Wells's dying crab on a red beach, and onward How to communicate across times greater than a millennium I treated in *Deep Time: How Humanity Communicates across Millennia* (New York: Avon, 1999), concluding that few deliberately left messages ever persist. The future will probably know little about us, for even our artifacts cannot endure. For example, Dyson shells constructed around stars could be technology's distant goal, and have inspired many science-fiction novels, but such ideas work only as long as stars burn, which means about one hundred billion years. One could look longer, and some did. In this spirit I edited *Far Futures* (New York: Tor, 1995), looking at the long view. Two of the five novella authors therein, Poul Anderson and Charles Sheffield, have died; alas, mortality versus the abyss. There are myriad other literary examples of writers and scientists confronting a truly ultimate question—of Last Things.

Chapter 14: "After the Stars Are Gone" by George Zebrowski

1. Fred Adam and Greg Laughlin, in their 1997 "Review of Modern Physics" paper, and later a profound book, *The Five Ages of the Universe* (New York: The Free Press, 1999).
2. Fred Adam and G. Laughlin, *The Five Ages of the Universe* (New York: The Free Press, 1999), xxviii.
3. Arthur C. Clarke, "The Long Twilight," in *Profiles of the Future,* Millennium Edition (London: Gollancz, 1999), 210.
4. C. M. Kornbluth, "The Failure of Science Fiction as Social Criticism," in *The Science Fiction Novel* (Chicago: Advent, 1959), 100.
5. Sir James Jeans, *The Mysterious Universe* (London: Macmillan, 1933).
6. Fred Adam and G. Laughlin, *The Five Ages of the Universe,* New York: The Free Press, 1999), xxviii.
7. Mileva Marić , letter to Einstein, cited in *Einstein: A Biography,* by Jürgen Neffe (New York: Farrar, Straus, and Giroux, 2007), 74.
8. This chapter is a sketch of George Zebrowski's commercially suppressed, unpublished novel *After the Stars Are Gone*, to which title and thematic details he now stakes his claim.

About the Contributors

Amara D. Angelica is editor of KurzweilAI.net and its daily *Accelerating Intelligence* newsletter. Her background includes more than thirty years as a technology/science journalist, inventor, aerospace engineer, radio producer, and musician.

Catherine Asaro is a quantum physicist with a Ph.D. in chemical physics from Harvard who has done research at the Max Planck Institut für Astrophysik in Germany, and the Harvard-Smithsonian Center for Astrophysics. Her fiction includes *The Quantum Rose*, which won the 2001 Nebula Award for best novel. A former ballerina, she established and heads Molecudyne Research, and also directs the nationally ranked math teams for the Howard Area Homeschoolers.

Gregory Benford holds a Ph.D. from the University of California, San Diego, and is a professor of physics in the department of physics and astronomy at the University of California, Irvine, a prize-winning fiction writer, author of *Deep Time* (1999), and, with his wife, Elisabeth Malartre, *Beyond Human: Living with Robots and Cyborgs* (2007).

Robert Bradbury studied at Harvard, has been a computer programmer, and for some years ran the Aeiveos Corporation doing R & D to understand and develop cures for aging. He invented the idea of Matrioshka Brains, the reshaped star systems designed for optimal energy use and maximal intelligence discussed in several chapters of this book.

Damien Broderick, Ph.D., a senior fellow in the School of Culture and Communication, University of Melbourne, is a critical theorist of the arts and sciences, a novelist, and a popular-science writer. His most recent novels are the diptych *Godplayers* (2005) and *K-Machines* (2006), and his most recent non-fiction book is *Outside the Gates of Science* (2007).

Sean M. Carroll, Ph.D., is a theoretical physicist at Caltech, specializing in cosmology, gravitation, and thermodynamics. He writes frequently at the lively blog Cosmic Variance, http://cosmicvariance.com/.

Anne Corwin is an electrical engineer specializing in electromagnetics, with a degree from California Polytechnic State University. She has a strong interest in radical life extension, and divides her time between research and writing in the areas of longevity, neurological diversity, technological trends, and ethics.

Dougal Dixon is a science writer whose best-known books attempt to present factual concepts emphasized and illustrated with fictitious examples: *After Man: A Zoology of the Future* (1981, on evolution), *The Future is Wild* (2002, on ecology), and the forthcoming *Greenworld* (human impact on natural systems).

Robin Hanson holds a Ph.D. from Caltech. After years of working in artificial intelligence, he is now an associate professor of economics at George Mason University, and a founder of the new field of prediction markets.

Steven B. Harris is a geriatrician and internist with long-standing interests in healthy longevity and cryonics. Dr. Harris is director of research at Critical Care Research, Inc., a company specializing in basic physiology research into rapid cooling (mild and moderate hypothermia) by cold fluorocarbon lung lavage, for treatment of postresuscitation syndrome and shock.

Jim Holt is a journalist who writes mostly on philosophical and scientific subjects. He splits his time between New York and Paris, and is a regular contributor to, among other publications, *The New Yorker*, *The New York Times Book Review*, *Slate*, *The New York Review of Books*, and *The American Scholar*.

Lisa Kaltenegger, who holds a Ph.D. in astrophysics, is a postdoctoral fellow at Harvard-Smithsonian Center for Astrophysics, specializing in extrasolar planet search and related science, simulations of detected planetary signals, biomarkers in the atmosphere, and evolution over geological time. She was named an American Young Innovator for 2007 by *Smithsonian* magazine.

Wil McCarthy is an engineer, novelist (notably with the *Collapsium* sequence, 2002-2005), science writer (*Hacking Matter*, 2003), columnist, and entrepreneur. Previously a flight controller for Lockheed Martin Space Launch Systems, and later an engineering manager for Omnitech Robotics and CTO of the aerospace research laboratory Galileo Shipyards, he is now president of The Programmable Matter Corporation.

Rudy Rucker holds a Ph.D. in mathematics from Rutgers, and is an emeritus professor of mathematics and computer science at San Jose State University, an expert on the fourth dimension and infinity, a novelist, and author of *The Lifebox, the Seashell, and the Soul* (2005). Some of the ideas in Dr. Rucker's essay are brought to life in his novel *Postsingular* (2007), and are discussed in his blog at www.rudyrucker.com.

Pamela Sargent is a prize-winning writer and anthologist whose novels have dealt presciently with such topics as human cloning (*Cloned Lives*, 1976), greatly extended longevity (*The Golden Space*, 1982), and terraforming (the *Venus* trilogy, 1986-2001).

George Zebrowski's degree in the history and philosophy of science and math has informed his career as a writer. He came to notice with his impressive novel *Macrolife* (1979), which follows a single family from the near future to the ends of time. Arthur C. Clarke called it "a worthy successor to Olaf Stapledon's *Star Maker*." His related, Kafkaesque novel about custodial space habitats, *Brute Orbits*, won the Campbell Memorial Award in 1999.